New Injection Fuels for Blast Furnace Ironmaking

Xu Runsheng Zhang Jianliang

Beijing
Metallurgical Industry Press
2023

Metallurgical Industry Press
39 Songzhuyuan North Alley, Dongcheng District, Beijing 100009, China

Copyright © Metallurgical Industry Press 2023. All rights reserved.

No part of this publication may be reproduced or transmitted in any form or by any means, electronic or mechanical, including photocopying, recording, or any information storage and retrieval system, without permission in writing from the copyright owner.

图书在版编目(CIP)数据

高炉喷吹燃料资源拓展与工业应用 = New Injection Fuels for Blast Furnace Ironmaking:英文/徐润生,张建良著.—北京:冶金工业出版社, 2023.7

ISBN 978-7-5024-9485-8

Ⅰ.①高… Ⅱ.①徐… ②张… Ⅲ.①高炉炼铁—燃料—研究—英文 Ⅳ.①TF526

中国国家版本馆 CIP 数据核字(2023)第 073866 号

New Injection Fuels for Blast Furnace Ironmaking

出版发行	冶金工业出版社	电 话	(010)64027926
地 址	北京市东城区嵩祝院北巷 39 号	邮 编	100009
网 址	www.mip1953.com	电子信箱	service@mip1953.com

责任编辑 卢 敏 张佳丽 美术编辑 彭子赫 版式设计 郑小利
责任校对 梅雨晴 责任印制 窦 唯

北京捷迅佳彩印刷有限公司印刷
2023 年 7 月第 1 版,2023 年 7 月第 1 次印刷
787mm×1092mm 1/16;15.5 印张;371 千字;235 页
定价 126.00 元

投稿电话 (010)64027932 投稿信箱 tougao@cnmip.com.cn
营销中心电话 (010)64044283
冶金工业出版社天猫旗舰店 yjgycbs.tmall.com
(本书如有印装质量问题,本社营销中心负责退换)

Introduction

Iron and steel industry is an important basic industry of China's national economy, which consumes a lot of energy and mainly relies on coal resources. However, the distribution of coal resources in China is extremely uneven. Among the proven reserves, bituminous coal accounts for 73.7%, anthracite accounts for 7.9%, lignite accounts for 6.8%, and other coals accounts for 11.6%, while the reserves of high-quality coking coal and fat coal in bituminous coal only account for 7.9%. Therefore, the resources of coking coal and fat coal on which the coke production in blast furnace depends, as well as the reserves of anthracite for blast furnace injection, are facing great challenges. Pulverized coal injection in blast furnace is an important measure for iron and steel enterprises to alleviate the limitation of coking coal resources, reduce coke ratio and pig iron production cost, and is the mainstream trend of global ironmaking technology development. However, 100% high quality anthracite injection can no longer meet the requirements of environmental protection, energy conservation, sustainable development of resources, cost reduction and efficiency improvement for blast furnace ironmaking.

More and more iron and steel enterprises continue to expand the scope of coal resources for blast furnace ironmaking. Meager-lean coal, bituminous coal and lignite coal are widely used for blast furnace injection, and semi-coke resources are tried to be used for blast furnace injection. Although some breakthroughs have been made, it faces a lot of technical problems. Among them, the flammability and explosiveness of high volatile bituminous coal bring serious safety problems to blast furnace injection. The frequent changes of various injection resources lead to the fluctuation of the hearth thermal state and gas flow, which seriously restricts the stability of blast furnace smelting. The main reason is that iron and steel enterprises lack fine management and scientific selection of new injection fuel, and fail to establish a systematic blast furnace injection process scheme according to the characteristics of new fuel.

In order to solve issues, with the support of major national and enterprise research projects, our team has systematically summarized the basic theory and industrial practice solutions in the process of injecting new fuels into blast furnaces through more than 20 years of interdisciplinary cooperation research and application. We have established a technical system for injecting unconventional fuels such as low rank coal, semi-coke, and biomass into blast furnace, and revealed the effect of blast furnace injection of natural gas, coke oven gas and other hydrogen fuel on blast furnace smelting and carbon reduction. To this end, we have initiated the writing of the book *New Injection Fuels for Blast Furnace Ironmaking*, with the aim of providing more scientific guidance for the scientific evaluation, economic selection, and efficient use of new fuel resources for blast furnace injection, thereby promoting energy conservation, emission reduction, and green development of the steel industry. The book has six chapters, including the overview of fuel resources for blast furnace injection, technology and industrial application of low rank coal injection into blast furnace, technology and industrial application of semi-coke injection into blast furnace, basic research on blast furnace biomass injection technology, basic research and industrial application of hydrogen-rich fuel injection into blast furnace, and prospects for the development of blast furnace injection fuel resources in China. This book focuses on how to systematically and scientifically evaluate the new fuel used in blast furnace injection and its efficient and safe application in industrial practice, and pays more attention to authority, systematization and practicability in the content, so as to make readers obtain a more systematic and comprehensive technical route for the expansion and efficient use of blast furnace fuel resources.

On the occasion of the publication of this book, I would like to express my special thanks to Professor Yang Tianjun of the University of Science and Technology Beijing for his guidance and suggestions on this work. Associate Professor Xu Runsheng undertook the planning, coordination and extensive editing work of this book, and put in hard work. Qi Chenglin, Hu Zhengwen, Wang Guangwei, Zheng Changle, Song Tengfei, Chai Yifan, Wang Haiyang, Wang Peng, Liu Siyuan, Lin Hao, Wei Guangyun, Li Rongpeng, and other graduate students have participated in the experimental work of the projects. Dang Han, Ye Lian, Zhang Cuiliu, Duan Sijiu, Bao Jian, Zhang Yuchen, Cao Minghui, Wang Shenyang, Zhao Peng, Hu Xiaotian give a lot of help in the picture drawing and English writing. In addition, these research works have been strongly supported by Shougang Corporation, Shenmu Lantan Group Co., Ltd., Jiuquan Iron & Steel (Group) Co., Ltd., Baotou Iron and Steel Company and other enterprises. We gratefully acknowledge the financial support from Ministry of Science and Technology of China and National Natural Science Foundation of China (52274316, 51704216, 2022YFE0208100, 2022A01003, ZDJB08, 202210700037, 2021ZXD01).

As scientific research itself is a process of continuous exploration and deepening, there are inevitably shortcomings in this book. We sincerely look forward to readers' correction.

Zhang Jianliang

July 1, 2023, University of Science and Technology Beijing

Contents

Chapter 1　Overview of Fuel Resources for Blast Furnace Injection ············ 1

1.1　Overview of the development of injection fuel for blast furnace ················ 1
1.2　Overview of the development of pulverized coal injection for blast furnace ············ 2
1.3　Chapter summary ·· 5
References ·· 5

Chapter 2　Technology and Industrial Application of Low Rank Coal Injection into Blast Furnace ·· 7

2.1　Metallurgical properties of bituminous coal and lignite ································ 7
　2.1.1　Proximate analysis ·· 7
　2.1.2　Ultimate analysis ·· 12
　2.1.3　Ash composition analysis ·· 16
　2.1.4　Ash melting point ·· 18
　2.1.5　Calorific value ·· 21
　2.1.6　Grindability ·· 28
　2.1.7　Adhesion ·· 30
　2.1.8　Ignition point ·· 31
　2.1.9　Explosibility ·· 33
　2.1.10　Flowability and injectability ·· 36
　2.1.11　Combustibility ·· 40
　2.1.12　Reactivity ·· 45
2.2　Interaction law of mixed injection of bituminous coal, lignite and anthracite ········ 47
　2.2.1　The influence of bituminous coal and lignite on the basic performance of blended coal ·· 47
　2.2.2　Effects of bituminous coal and lignite on ignition point and explosiveness of mixed coal ·· 48
　2.2.3　Effect of bituminous coal and lignite on combustibility of blended coal ········ 49
　2.2.4　Effects of bituminous coal and lignite on gasification reactivity of blended coal ·· 50
　2.2.5　Research on optimization of coal blending scheme ···························· 51
2.3　Effect of particle size on combustion performance of low-rank coal ··············· 52
　2.3.1　Effect of particle size on combustibility of anthracite ························· 52

2.3.2	Effect of particle size on the combustibility of bituminous coal	53
2.3.3	Effect of particle size on combustibility of lignite	54
2.3.4	Combustion characteristics and kinetic analysis of pulverized coal with different particle size	55
2.3.5	Effect of particle size on combustibility of mixed coal	62
2.4	Full bituminous coal injection technology and industrial application	63
2.4.1	Safety assessment and rectification of blast furnace coal injection system	63
2.4.2	Operating rules for blast furnace full injection of bituminous coal	66
2.4.3	Industrial application practice of blast furnace full injection of bituminous coal	68
2.5	High proportion lignite injection technology and industrial application	73
2.5.1	Current situation of coal injection in blast furnace of lingsteel	74
2.5.2	Industrial experiment plan design	74
2.5.3	Industrial experiment results analysis	76
2.6	Chapter summary	78
	References	79

Chapter 3 Technology and Industrial Application of Semi-Coke Injection into Blast Furnace ... 81

3.1	Overview of semi-coke	81
3.2	Metallurgical properties of semi-coke	82
3.2.1	Proximate analysis of semi-coke	82
3.2.2	Ultimate analysis of semi-coke	84
3.2.3	Ash composition analysis	85
3.2.4	Ash melting point	86
3.2.5	Heating value	88
3.2.6	Grindability	89
3.2.7	Adhesion	90
3.2.8	Ignition point and explosivity	91
3.2.9	Flowability and injectability	92
3.2.10	Combustibility	93
3.2.11	Gasification characteristic	96
3.3	Combustion characteristics and mechanism of semi-coke mixed with injected coal	98
3.3.1	Experimental samples and methods	98
3.3.2	TG/DTG curve analysis of combustion of single semi-coke and single bituminous coal	99
3.3.3	TG/DTG curve analysis of mixed combustion of semi-coke and bituminous coal in different proportions	100

3.3.4	Combustion characteristic index analysis.	101
3.3.5	Mechanism analysis of factors affecting combustion characteristics	103

3.4 Effect of semi-coke addition on metallurgical properties of mixed coal ········· 106
 3.4.1 Effect of semi-coke addition on safety performance of mixed coal ········· 106
 3.4.2 Influence of adding semi-coke on the reactivity of mixed coal gasification ······ 107
3.5 Regulation of grindability of semi-coke and optimization of injection size ········· 110
 3.5.1 Control technology for grindability of semi-coke ········· 110
 3.5.2 Optimization of mixing size of semi-coke and injection coal ········· 111
3.6 Mixed injection technology and industrial application of semi-coke and pulverized coal ········· 115
 3.6.1 Current situation of coal injection into the blast furnace of Xinxing Ductile Iron Pipes Co., Ltd. ········· 115
 3.6.2 Determination of industrial experiment plan ········· 116
 3.6.3 Comparative analysis of blast furnace smelting parameters ········· 118
3.7 Analysis of energy saving and emission reduction effect of semi-coke injection ········· 123
 3.7.1 Energy consumption of ironmaking process ········· 123
 3.7.2 CO_2 emission of ironmaking process ········· 124
 3.7.3 Effect of semi-coke powder injection on energy saving and emission reduction of blast furnace ········· 125
3.8 Chapter summary ········· 130
References ········· 131

Chapter 4 Basic Research on Blast Furnace Biomass Injection Technology ········· 133

4.1 Overview of biomass ········· 133
4.2 Study on physicochemical properties of biomass char ········· 134
 4.2.1 Experimental materials and preparation ········· 134
 4.2.2 Biomass char yield ········· 135
 4.2.3 Proximate and ultimate analysis ········· 136
 4.2.4 Microscopic morphology analysis ········· 136
 4.2.5 Specific surface area and pore structure analysis ········· 139
 4.2.6 Microcrystalline structure analysis ········· 141
4.3 Study on combustion characteristics of biomass char/pulverized coal in case of mixed injection ········· 144
 4.3.1 Experimental materials and methods ········· 145
 4.3.2 Mixed combustion characteristics and synergistic effect of biomass char/pulverized coal ········· 146
 4.3.3 Kinetic analysis of biomass char/pulverized coal mixed combustion ········· 161
4.4 Effect of unburned residual carbon from biomass semi-coke on properties of

blast furnace coke and slag ········· 166
    4.4.1 Effect of unburned residual carbon on coke gasification reaction ········· 167
    4.4.2 Effect of unburned residual carbon on viscosity of blast furnace slag ········· 177
  4.5 Study on the changes in smelting parameters of mixed injection of biomass char and pulverized coal into blast furnace ········· 185
    4.5.1 Establishment of material and heat balance model for blast furnace smelting ········· 186
    4.5.2 Changes in smelting parameters of blast furnace mixed injection biomass semi-coke and pulverized coal ········· 193
  4.6 Chapter summary ········· 199
  References ········· 200

Chapter 5 Basic Research and Industrial Application of Hydrogen-Rich Fuel Injection into Blast Furnace ········· 203

  5.1 Brief introduction of blast furnace injection of hydrogen-rich fuel ········· 203
  5.2 Mixed injection of natural gas and pulverized coal into blast furnace ········· 205
    5.2.1 Overview of natural gas injection into blast furnace ········· 205
    5.2.2 Impact of mixed injection of natural gas and coal on the utilization coefficient of blast furnace ········· 206
    5.2.3 Effect of mixed injection of natural gas and coal on slag performance ········· 207
    5.2.4 Effect of mixed injection of natural gas and coal on operating parameters of blast furnace ········· 208
    5.2.5 Impact of natural gas injection rate on the coal combustion process ········· 209
    5.2.6 Impact of natural gas injection rate and coal ratio on the coal combustion process ········· 214
    5.2.7 The effect of mixed injection of natural gas and pulverized coal into blast furnace on the volume of gas in the bosh ········· 217
  5.3 Injection of coke oven gas into blast furnace ········· 219
    5.3.1 Significance and advantages of coke oven gas injection into blast furnace ········· 219
    5.3.2 Development history of coke oven gas injection into blast furnaces at home and abroad ········· 220
    5.3.3 Process flow of coke oven gas injection into blast furnace ········· 221
    5.3.4 Effect of coke oven gas injection on the smelting law of blast furnace ········· 223
  5.4 Chapter summary ········· 232
  References ········· 232

Chapter 6 Prospects for the Development of Blast Furnace Injection Fuel Resources in China ········· 233

Chapter 1 Overview of Fuel Resources for Blast Furnace Injection

The steel industry is the backbone of national development, providing an important raw material guarantee for China's economic and social development; at the same time, the steel industry is also a high energy-consuming and high-polluting industry, facing huge pressure of energy conservation and emission reduction today with the emphasis on ecological civilization construction. From the view of the whole steel smelting process, ironmaking is one of the most important parts of steel production, and is also one of the most energy-consuming and pollutant-emitting parts. Therefore, the research and development of new technology for blast furnace ironmaking has become an important way to save energy and reduce consumption, reduce CO_2 emissions and improve the competitiveness of the iron and steel industry. With the upsizing of blast furnace and the increase of pig iron output, the demand for metallurgical coke in the smelting process also increases gradually, and the consumption of coking coal resources also increases. However, coking coal is a non-renewable resource and is becoming increasingly scarce. Moreover, the environmental pollution caused by the coking process is very serious, so it is an inevitable trend to find a new fuel to replace part of the coke in the ironmaking industry. Blast furnace fuel injection technology is an innovative technology that can effectively reduce the coke consumption of blast furnace, and it can replace some coke to provide heat, reducing agent and carburizing agent to the blast furnace, thus reducing the coke consumption.

1.1 Overview of the development of injection fuel for blast furnace

As early as the 19th century, the idea of blast furnace fuel injection was proposed in Europe and the United States, and the relevant patents were applied for, but it was not until the middle of the 20th century that the application was gradually realized in industry. In 1947, the French Navi-Maison plant carried out industrial tests of fuel oil injection into blast furnaces. In 1948, the Soviet Union Terzhinsky plant tried to blow coal dust into blast furnaces. In 1957, Petrovsky plant in the Soviet Union began to blow natural gas into the blast furnace. Since then, countries around the world according to their own resource conditions and fuel prices choose to blow different fuels, for example, Russia and the United States are rich in natural gas resources, so a large number of natural gas injection. In the 1960s oil was cheap in world markets, heavy oil was injected into blast furnaces in large quantities all over the world; in the late '70s, due to high oil prices, most countries stop injecting oil into blast furnaces and gradually switch to inject pulverized coal. In 1990, 2/3 of the blast furnaces in Japan and Germany injected pulverized coal, injection rate was generally 50-80kg/tHM, by 1998, some blast furnaces had been injecting more than 200kg/

tHM. China started oil injecting into blast furnaces in the late 1950s, and most of the blast furnaces were oil injected in the early 1960s. In 1964, Capital Iron and Steel Company [1-2] and Anshan Iron and Steel Company achieved success in blast furnace injecting anthracite coal, and in 1966 a blast furnace in Shougang injects an average of 159kg/tHM of coal throughout the year. In addition to this, Chongqing Iron and Steel Company injected natural gas in the 1960s, during which some iron and steel companies also injected tar and asphalt, and gradually switched to pulverized coal injecting at the end of the 1960s.

1.2 Overview of the development of pulverized coal injection for blast furnace

Pulverized coal is the most important fuel for blast furnace injecting. The development of blast furnace coal injection can be traced back to 1840-1850. In 1840s, Banks proposed the idea of replacing part of the coke with anthracite, and between 1840 to 1845, the world's first industrial application was realized and patented in the ironworks of the Haute-Marne department in Boulogne, France. However, because of problems in the process, this technology was not popularized, and for more than 100 years afterwards, the blast furnace coal injection technology developed very slowly and made little progress until the early 1960s, when it began to develop rapidly.

The first large-scale blast furnace coal injection test in industry was completed by the International Iron and Steel Union in 1961 at Hanna's No. 2 blast furnace in North America. In the early 1960s, many countries in the world, such as France, the United Kingdom, the United States, the Soviet Union and Germany, began a large number of blast furnace coal injection tests and industrial researches.

Blast furnace coal injection trials began in the United States in 1962, and industrial applications officially began in 1966 at Amco Steel's Ashland plant on the Bellefonte blast furnace using a injection system developed in cooperation with Babcock-Wilkes, which reached an annual average coal injection stiffness of about 45kg/tHM in the late 1960s[3]. After the technical improvements, the average coal injection of the Amada blast furnace reached a maximum of 58kg/tHM in 1973. The coal injection tests in the United States were conducted at the same time that similar tests were conducted in Australia, Europe and Asia.

When the initial success of blast furnace coal injection test, blast furnace injecting heavy oil and natural gas technology because of the advantages of simple process, low investment, etc. had been more development and application. However, in the 1970s, due to the emergence of the two oil crises, ironworks around the world began to widely apply coal injection technology again, accelerating the research and development of coal injection technology, especially in Japan and Europe in the practical application of a major breakthrough. After entering the 1980s, coal injection technology was really widely used and developed in the world.

Japan began to apply blast furnace coal injection technology in 1981, and the proportion of blast furnaces using coal injection technology reached 81.8% by the end of 1992, and all blast furnaces in Japan used blast furnace coal injection technology in 1995, and the monthly average coal injection ratio of many blast furnaces exceeded 200kg/tHM.

In the early 1990s, the rapid development of blast furnace coal injection technology, coal ratio increased significantly, some manufacturers in Western Europe, Japan and other countries relied on their own good technical equipment and research and development capabilities to achieve the annual coal injection ratio of more than 200kg/tHM record, and has remained the world's leading level[4], Japan's Kakogawa No. 1 blast furnace, Kobe No. 3 blast furnace, Fukuyama No. 4 and Juntsu No. 3 blast furnace coal injection ratio were greater than 200kg/tHM, of which Kakogawa No. 1 blast furnace and Kobe No. 3 blast furnace lasted 13 months and 12 months, respectively. At the same time, the United States, Italy, the United Kingdom and other countries were also vigorously developing the blast furnace coal injection technology, the United States of America Liancoli No. 13 blast furnace reached 210kg/tHM coal injection ratio for 2 months, Italy's Taranto No. 4 blast furnace reached 204kg/tHM coal injection ratio for 2 months, the United Kingdom's Queen Victoria blast furnace coal injection ratio also exceeded 200kg/tHM, and maintained for 2 months [5].

Due to the different resource status of each country, the blast furnace injecting fuel also varies, and pulverized coal was the main injecting fuel for the blast furnace. North American blast furnaces were mostly injection with oil and natural gas in addition to pulverized coal. In order to reduce production costs, foreign steel companies attached great importance to increasing the coal injection ratio while maintaining a low fuel ratio. In 2011, the average coal injection ratio of major blast furnaces in European countries was 143kg/tHM, and the average coal injection ratios of Tata Steel Elmoiden No. 6 and No. 7 blast furnaces were both above 200kg/tHM, which were at a relatively leading level[6], and the coal injection ratios of three European blast furnaces in 2014: No. 3 blast furnace in Lulea Sweden, No. 2 blast furnace in Thyssen Schwergen, Germany, and No. 8 blast furnace in Thyssenheimborn, Germany, were 144kg/tHM, 174kg/tHM, and 198kg/tHM, respectively, with the highest fuel ratio of 502kg/tHM. In 2013, the average coal injection ratio of Pohang Steel Pohang Plant in Korea was 179kg/tHM, which reached a high level. In 2015, most of the indicators of Korean blast furnaces were more advanced, with fuel ratio below 500kg/tHM, the average coal injection ratio of Gwangyang and Pohang were 170.6kg/tHM and 170.5kg/tHM respectively. The average coal injection ratio of Japanese blast furnaces increased year by year from 2008 to 2013, and in 2011, the average coal injection ratio of Japanese blast furnaces was 151kg/tHM, and the average coal injection ratio of Nippon Steel & Sumikin Nagoya No. 3 blast furnace reached 189kg/tHM, and in 2013, the average coal injection ratio of Japanese blast furnaces reached 169kg/tHM with low fuel ratio [7].

The history of blast furnace coal injection in China began in the 1960s, when Shougang and Ansteel conducted research on blast furnace coal injection technology and put it into production[8]. Therefore, China was one of the first countries to realize blast furnace coal injection technology. But since then for more than 20 years, the blast furnace blasting was basically anthracite, and even a few blast furnace blasting raw coal without washing, and by the backward coal blasting equipment and other issues, China's blast furnace coal injection process development was slow, the average coal injection ratio of only 50-60kg/tHM.

After the 1990s, blast furnace coal injection technology was included in the national scientific and technological research program, large blast furnaces were all equipped with coal injection devices, while the number of blast furnaces with pulverized coal injection was increasing, and large coal injection had become an important initiative in the development of China's blast furnace ironmaking technology. Since 1995, the coal injection ratio of blast furnaces in China had been increasing, and the average coal injection ratio of key enterprises was only 58kg/tHM in 1995, which had reached 118kg/tHM by the end of the 20th century and 125kg/tHM in 2002[9]. Entering the 21st century, due to the significant increase in world oil prices, coupled with the current situation of China's more coal and less oil resources, making the blast furnace coal injection benefits significantly increased, China's blast furnace coal injection technology stepped to a new stage of rapid development, the chinese new or proposed blast furnace design coal injection ratio is mostly above 200kg/tHM [10].

At present, all blast furnaces in China are implemented to pulverized coal injection, Baosteel, Shougang, Ansteel and other technologically advanced enterprises had approached or reached the world's advanced level, for example, Baosteel No.1 blast furnace coal ratio was stable above 200kg/tHM, of which, from 1999 to 2002, the coal ratio was stable at 230-240kg/tHM [11], large and medium-sized iron and steel enterprises continued to expand the coal injection machinery. In recent years, the international advanced level of blast furnace fuel ratio is less than 500kg/tHM, and only Baosteel Group's blast furnace fuel ratio reached this level in 2017. The blast furnace coal injection example of China's key enterprises in the past years as shown in Fig. 1-1 by the statistics of China Steel Association [12]. In recent years, the overall level of China's blast furnace coal injection ratio is not high, ironmaking enterprises because the price difference between coke and pulverized coal is shrinking, no longer pursue too high coal injection ratio, but seek economic coal injection ratio, improve the replacement ratio of pulverized coal, to achieve the optimization of ironmaking costs.

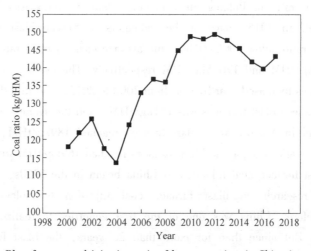

Fig. 1-1 Blast furnace coal injection ratio of key enterprises in China in recent years

1.3 Chapter summary

China is one of the early countries in the world to achieve coal injection into blast furnaces, and also one of the countries that have been insisting on coal dust injection. The popularity of China's blast furnace coal injection is high, the total amount of coal injection is far ahead in the world, but due to the geographical differences, the level of raw materials and the physical properties and process performance of the coal used for injection are also different from plant to plant, the level of blast furnace automation and operation level is also different, most manufacturers are affected by a number of factors such as the weak research force house and imperfect detection means, resulting in a wide range of fluctuations in the blast furnace coal injection amount of the Chinese blast furnace ironmaking, about 100-200kg/tHM.

Coal accounts for 70% of the energy structure of China's steel production, especially the energy required for blast furnace iron making is 70% to 80% from carbon combustion. Although China's coal reserves are large, with total coal reserves of 800 billion tons, the proportion of each coal type varies greatly and is extremely unevenly distributed. China's annual blast furnace injection coal powder more than 100 million tons, injection coal is mainly anthracite, with a small amount of bituminous coal. China's anthracite resources are scarce, while low-rank coal resources are abundant, and there are obvious contradictions between the injection structure and resource structure. Therefore, expanding the range of blast furnace injection fuel resources and relieving the dependence of China's blast furnace injection coal on high-quality anthracite resources is an urgent task for ironworkers at present.

References

[1] Liu Yuncai. Blast furnace coal injection limit day practice and development prospect [C]// Proceedings of the 1993 Annual Ironmaking Academic Conference of the Chinese Society of Metals. 1993 Annual Ironmaking Academic Conference, Tangshan, Hebei, 1993: 718-724.

[2] Liao Xizhen, Yu Lixing, Yan Wei. Blast furnace blown bituminous coal [J]. Iron and Steel, 1979 (4): 11-17.

[3] Maki A, Sakai A, Takaqaki N, et al. High rate coal injection of 218kg/t at Fukuyama No. 4 blast furnace [J]. ISIJ International, 1996, 36 (6): 650-657.

[4] Yang Tianjun, Liu Yingshu, Yang Mian. Blast furnace oxygen-rich coal injection-oxygen-coal mixing and combustion [M]. Beijing: Science and Technology Press, 1998.

[5] Wang Wei, Bi Xuegong, Fu Shimin. Current status and main technical measures of blast furnace coal injection at home and abroad [J]. Journal of Wuhan University of Science and Technology (Natural Science Edition), 2002, 21 (1): 11-12.

[6] Naito M, Takeda K, Matsui Y. Ironmaking technology for the last 100 years: Deployment to advanced technologies from introduction of technological know-how, and evolution to next-generation process [J]. ISIJ International, 2015, 55 (1): 7-35.

[7] Wang Weixing. Analysis of factors influencing blast furnace fuel ratio [N]. World Metal Herald, 2016-08-02 (B02).

[8] Liu Yuncai. Achievement and development of blast furnace coal injection in China [J]. Iron and Steel, 1994, 29 (9): 71-76.

[9] Pei Xiping. Analysis of the current situation and supply and demand trend of blast furnace coal injection in China [J]. Ironmaking, 2003, 22 (5): 33-36.

[10] Hou Xing. Status and development of blast furnace coal injection at home and abroad [J]. Jiangxi Metallurgy, 2012 (3): 40-42.

[11] Xu Wanren, Li Zhaoyi, Guo Yanling. Production practice of economic coal injection ratio in Baosteel No. 1 blast furnace [J]. Ironmaking, 2010, 29 (1): 29-31.

[12] Wang Weixing. Review of iron production technology in China in 2017 [N]. World Metal Herald, 2018-03-13 (B02).

Chapter 2 Technology and Industrial Application of Low Rank Coal Injection into Blast Furnace

At present, anthracite is the main coal used in blast furnace ironmaking, and a few iron and steel enterprises use meagre coal, meager lean coal, lean coal, not caking coal and anthracite mixed injection. The main reason that anthracite occupies the dominant position in injection fuel for a long time is that especially anthracite has the advantages of high heat value, good injection safety, high coal coke replacement ratio and easy reception in blast furnace. However, from the distribution of coal resources, our anthracite resources are less than 8%, the reserves of bituminous coal are about 73.7%, low rank coal resources reserves are abundant[1]. At the same time, from the economic perspective, bituminous coal price is especially high, low level coal price is low. Therefore, both from the perspective of sustainable development of resources and from the perspective of cost reduction and efficiency increase in blast furnace ironmaking, it is necessary to enhance the proportion of low rank coal used in blast furnace injection and develop low rank coal injection technology. The low rank coal discussed in this book is mainly represented by lignite and high-volatile bituminous coal.

2.1 Metallurgical properties of bituminous coal and lignite

Basic characteristics of blast furnace coal injection refer to the intrinsic characteristics of pulverized coal, usually including proximate analysis, elemental analysis, ash composition, ash melting point, heating value and adhesion. The technical performance of blast furnace coal injection refers to the performance that should be possessed in order to meet the needs of the blast furnace injection process. The performance index is used to balance the safety and stability of the coal injection in the blast furnace pulverization process, pulverized coal transportation process and combustion and heat release process, usually including grindability, fluidity, jet property, combustion property, reactivity, ignition point and explosive.

2.1.1 Proximate analysis

The proximate analysis of coal is a general term for the determination of four analytical items, including moisture (M), ash (A), volatile (V) and fixed carbon (FC). The proximate analysis of coal is the main index and the basic basis of evaluating coal quality. Usually, the moisture, ash and volatile of coal are measured directly, while the fixed carbon is calculated by difference subtraction. Iron and steel enterprises in the proximate analysis of coal injection generally according to the national standard GB/T 212—2008.

(1) Ash content. Ash content refers to all kinds of non-combustible minerals in coal,

including inert and non-inert minerals. Among them, non-inert minerals refer to chemical changes that can occur through catalytic or non-catalytic reactions. The main components of ash in coal after coal combustion are SiO_2, Al_2O_3, Fe_2O_3, CaO, MgO, TiO_2, Na_2O, etc. Blast furnace injection requires the ash content of coal, as low as possible. Because with the increase of ash content in pulverized coal, the theoretical combustion temperature and pulverized coal combustion efficiency of blast furnace will decrease, resulting in the reduction of coal coke replacement ratio. The influence of pulverized coal ash content on coke ratio of blast furnace is equivalent to that of coke ash content on coke ratio. In actual production, it is also found that the influence of pulverized coal ash injection in blast furnace on coke ratio has different degrees of linear relationship, as shown in Fig. 2-1.

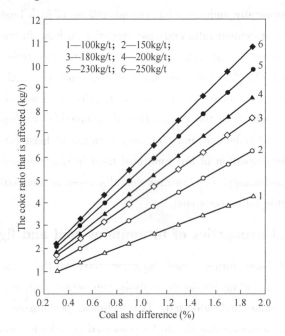

Fig. 2-1 Effect of pulverized coal ash on coke ratio

As can be seen from Fig. 2-1, when the coal ratio of 150kg/t is set in the blast furnace, the actual coke ratio will be reduced by nearly 3% for every 1% increase in pulverized coal ash content, which is a great influence on the production of the blast furnace. Therefore, coal with low ash content should be selected as far as possible. The ash content standard of blast furnace coal injection is shown in Table 2-1.

Table 2-1 Coal ash classification

Serial number	Level name	Code	A_d (%)
1	super low ash coal	SLA	5.00
2	low ash coal	LA	5.01-10.00
3	low medium ash coal	LMA	10.01-20.00
4	medium ash coal	MA	20.01-30.00

Continued Table 2-1

Serial number	Level name	Code	A_d (%)
5	medium high ash coal	MHA	30.01-40.00
6	high ash coal	HA	40.00-50.00

(2) Volatile matter. Volatiles are gaseous products (except water vapor) that escape from coal under the condition of isolation of air heating. The volatiles content of coal decreases with the increase of coalification degree. The volatile components of coal mainly include CO_2, CO, H_2, CH_4, C_2H_2, C_3H_3, C_3H_6 and a small amount of cyclic hydrocarbon (C_mH_n).

The volatile content of coal has great influence on blast furnace injection effect. It is generally believed that the higher the volatile content, the higher the combustion rate of coal, from this point of view, the higher the volatile content of coal is more favorable. However, due to the high volatile content, the explosive performance of coal is enhanced, coal burning in front of the tuyere needs to absorb more heat, so more heat compensation is required. The volatile component standard of blast furnace coal injection is shown in Table 2-2.

Table 2-2 Volatile classification of coal

Serial number	Level name	Code	V_{daf}(%)
1	super low volatile coal	SLV	≤10.00
2	low volatile coal	LV	>10.00-20.00
3	medium volatile coal	MV	>20.00-28.00
4	medium high volatile coal	MHV	>28.00-37.00
5	high volatile coal	HV	>37.00-50.00
6	super high volatile coal	SHV	>50.00

(3) Fixed carbon. Fixed carbon, the main heating part of coal, is the combustible material after removing volatiles and ash. In a certain range, the higher the fixed carbon, the greater the heating value, the more favorable for coal injection. Therefore, the fixed carbon of coal for injection is generally required to be between 70% and 87% (research shows that when the fixed carbon is greater than 87%, the heating value decreases with the increase of fixed carbon).

(4) The coal used in blast furnace injection is mostly washed coal, the total moisture content is generally higher, so that the coal is easy to break in the process of storage and transportation, increase the transportation cost, and reduce the calorific value of coal, heat consumption in the grinding of coal, reduce the output of coal mill. In addition, the water injected into the blast furnace before the tuyere to decompose heat absorption, increase compensation heat without compensation means will reduce the amount of injection, so the lower the total moisture, the better. The national standard stipulates that the total moisture of anthracite for blast furnace injection is no more than 12.00%; meagre coal, meagre lean coal, gas coal ($G_{R,I}$<50) total moisture ≤8.00%; total moisture of long flame coal, non-sticky coal and weak sticky coal is no

more than 14.00%.

The coal used in blast furnace injection in domestic iron and steel enterprises is shown in Table 2-3. Types of injection coal include anthracite, lean coal, bituminous coal, gas coal and lignite. From the point of view of statistics, iron and steel enterprises are mainly anthracite, anthracite types, and its composition fluctuation is also obvious, fixed carbon fluctuates between 64%-82%, volatile content fluctuates between 5%-15%, ash fluctuates between 8%-17%, moisture fluctuation between 0.5%-10%. General iron and steel enterprises choosing high quality anthracite usually use the "double seven" index, that is, the fixed carbon content is greater than 70%, calorific value is greater than 7000 CAL.

Table 2-3 Analysis of coal industry in domestic iron and steel enterprises

Serial number	Species	Coal	Moisture (%)	Volatile (%)	Ash (%)	Fixed carbon (%)
1	anthracite	Daiwang	3.95	6.91	11.58	77.56
2	anthracite	Qingding 1	3.7	10.63	11.28	74.39
3	anthracite	Baiyangshu	5.77	11.31	13.12	69.8
4	anthracite	Fengshan	8.23	14.21	12.99	64.57
5	anthracite	Jiaozuobei	5.27	12.41	13.11	69.21
6	anthracite	Huaren	5.7	12.21	13.15	68.94
7	anthracite	Xinjing	9.97	12.85	13.18	64
8	anthracite	North Korea	6.05	6.91	16.43	70.61
9	anthracite	Jingxing	6.37	13.45	13.64	66.54
10	anthracite	Botou	6.02	9.6	13.33	70.87
11	anthracite	Inner Mongolia	3.34	9.13	12.96	74.57
12	anthracite	Gaoping	1.98	9.1	13.88	75.04
13	anthracite	Haolin	0.5	9.69	10.61	79.2
14	anthracite	Baiyangshu	0.58	7.28	10.2	81.94
15	anthracite	Russian coal	0.72	5.06	16.58	77.64
16	anthracite	Hengyuan	1.86	9.01	9	80.13
17	anthracite	Qingding 2	2.1	7.78	10.55	79.57
18	anthracite	Shenhuo	1.99	7.84	11.07	79.1
19	anthracite	Dongpei	5.25	6.91	12.82	75.02
20	anthracite	Guanghui	5.67	9.42	16.74	68.18

Continued Table 2-3

Serial number	Species	Coal	Moisture (%)	Volatile (%)	Ash (%)	Fixed carbon (%)
21	anthracite	Ning coal	1.61	9.73	14.08	74.58
22	anthracite	Zhongxin	4.02	9.8	8.52	77.66
23	lean coal	Sangeicun	3.37	13.48	11.02	72.13
24	lean coal	Lu'an	3.58	11.49	10.35	74.58
25	lean coal	electric cleaning coal of Ling Steel	0.2	14.19	10.06	75.55
26	bituminous coal	Shentong	5.72	33.77	7.68	52.83
27	bituminous coal	Xuanhua	5.54	33.5	9.7	51.26
28	bituminous coal	Gonghong	1.1	31.18	8.58	59.14
29	bituminous coal	Shenhua	5.27	23.94	9.04	61.75
30	bituminous coal	Steam coal	4.95	23.4	10.41	61.24
31	bituminous coal	bituminous coal of Ling Steel	7.6	26.48	11.8	54.12
32	gas coal	Jiyang gas coal	12.26	32.11	7.95	47.68
33	lignite	Jinkai	14.68	41.08	8.87	35.37
34	lignite	lignite of Ling Steel	13.35	33.84	11.43	41.38
35	lignite	Xinneng	18.13	36.2	7.08	38.59
36	lignite	Binxin	17.28	37.21	7.74	37.77
37	lignite	Guoleizhuang	18.55	31.67	10.88	38.9

The proximate analysis results of bituminous coal, lignite coal and anthracite are compared, as shown in Fig. 2-2 and Fig. 2-3. The composition of bituminous coal and lignite is significantly different from that of anthracite, that is, the volatile content of bituminous coal and lignite is 20%-30% higher than that of anthracite, and the fixed carbon of bituminous coal is 30%-40% lower than that of anthracite. It can be inferred that the effective heating value of lignite and bituminous coal is lower than that of anthracite in the process of injection. In addition, lignite is notable for its high water content, which is 13%-19%. Lignite and bituminous coal used in blast furnace injection belong to medium-high volatile and low-ash coal, while anthracite belongs to low-volatile and low-medium ash coal.

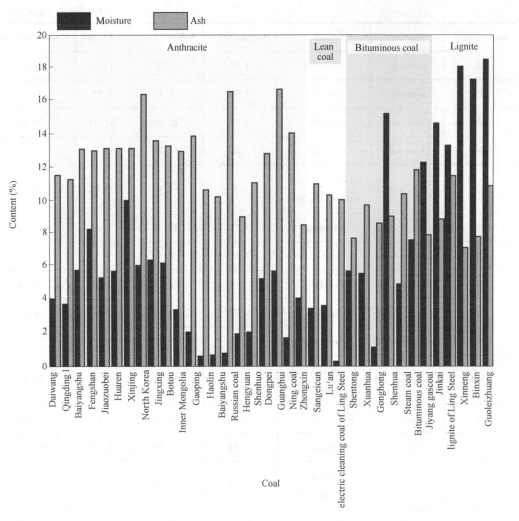

Fig. 2-2 Comparison of moisture and ash content in different coals

2.1.2 Ultimate analysis

Ultimate analysis of coal refers to the determination of elemental content of organic matter in coal. The ultimate analysis of coal mainly includes carbon, hydrogen, oxygen, nitrogen, sulfur and phosphorus. Among them, carbon, hydrogen and nitrogen are determined, oxygen is calculated, sulfur and phosphorus are determined separately according to the requirements of coal quality research and production. By measuring the content of main elements in coal, we can preliminarily understand the chemical properties of coal and the degree of coalification, calculate the calorific value of coal, estimate and predict the coking products of coal, low temperature retorting yield, and provide the basis for effective utilization. The ultimate analysis of coal is generally carried out in accordance with GB/T 31391—2015. The S, P and alkali metals in coal are harmful elements to blast furnace smelting, and the lower the content, the better. The sulfur content of the coal should be the same as the sulfur content of the coke used, generally requiring $S_{t,d}<1.0\%$. The

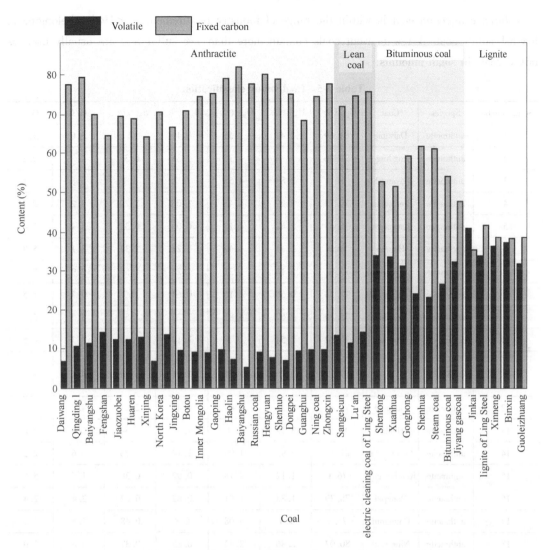

Fig. 2-3 Comparison of volatile and fixed carbon content in different coals

sulfur content standard of blast furnace coal injection is shown in Table 2-4.

Table 2-4 Coal sulfur classification

Seral number	Level name	Code	$S_{t,d}$ (%)
1	super low sulfur coal	SLS	≤0.50
2	low sulfur coal	LS	0.51-1.00
3	low medium sulfur coal	LMS	1.01-1.50
4	medium sulfur coal	MS	1.51-2.00
5	medium high sulfur coal	MHS	2.01-3.00
6	high sulfur coal	HS	>3.00

Table 2-5 shows the results of ultimate analysis of coal used in blast furnace injection in domestic iron and steel enterprises. It can be seen from the table that the sulfur content of most blast furnace injection coal is within the range of industry requirements (<1%), belonging to low ash sulfur coal. A few blowout coals contain more than 1% sulfur and are usually used as mixed coals in small amounts.

Table 2-5 Coal sulfur classification

Serial number	Species	Coal	C_{ad} (%)	H_{ad} (%)	O_{ad} (%)	N_{ad} (%)	S_{ad} (%)	H/C	O/C
1	anthracite	Daiwang	81.69	2.47	1.55	0.11	1	3.0	1.9
2	anthracite	Qingding 1	78.26	3.33	1.75	1.38	0.33	4.3	2.2
3	anthracite	Baiyangshu	74.3	2.83	4.55	0.1	1.7	3.8	6.1
4	anthracite	Fengshan	70.72	2.29	6.45	1.08	1.29	3.2	9.1
5	anthracite	Jiaozuobei	76.45	2.62	4.13	0.06	0.32	3.4	5.4
6	anthracite	Huaren	76.6	2.41	4.42	1.21	1.37	3.1	5.8
7	anthracite	Xinjing	75.01	2.16	5.31	1.23	1.56	2.9	7.1
8	anthracite	North Korea	77.53	0.58	1.16	0.64	0.2	0.7	1.5
9	anthracite	Jingxing	75.39	2.67	4.81	1.22	1.54	3.5	6.4
10	anthracite	Botou	76.32	2.54	3.5	1.17	2.11	3.3	4.6
11	anthracite	Inner Mongolia	79.17	3.45	3.51	1.01	0.98	4.4	4.4
12	anthracite	Gaoping	76.85	3.26	3.67	1.41	2.91	4.2	4.8
13	anthracite	Haolin	81.5	3.12	2.1	1.02	0.54	3.8	2.6
14	anthracite	Baiyangshu	81.12	2.95	1.99	1.14	0.76	3.6	2.5
15	anthracite	Russian coal	76.1	1.12	4.18	0.92	0.38	1.5	5.5
16	anthracite	Dongpei	78.39	1.88	1.91	0.82	0.62	2.4	2.4
17	anthracite	Gaunghui	75.3	2.1	3.08	0.8	0.48	2.8	4.1
18	anthracite	Ning coal	80.07	2.86	2.42	0.85	0.37	3.6	3.0
19	anthracite	Zhongxin	83.92	2.02	2.77	0.92	0.35	2.4	3.3
20	lean coal	Sangeicun	74.97	3.36	4.18	1.35	1.68	4.5	5.6
21	lean coal	Lu'an	80.97	3.49	2.08	0.09	0.14	4.3	2.6
22	bituminous coal	Shentong	67.25	3.29	12.12	0.76	0.54	4.9	8.0
23	bituminous coal	Xuanhua	69.15	3.43	10.72	0.78	0.35	5.0	15.5
24	bituminous coal	Gonghong	71.41	3.7	14.21	0.86	0.14	5.2	19.9
25	gas coal	Jiyang gas coal	75.37	4.47	8.18	1.69	1.16	5.9	10.9

Continued Table 2-5

Serial number	Species	Coal	C_{ad} (%)	H_{ad} (%)	O_{ad} (%)	N_{ad} (%)	S_{ad} (%)	H/C	O/C
26	lignite	Jinkai	60.76	2.76	13.87	0.64	0.19	4.5	22.8
27	lignite	Xinneng	66.58	3.82	19.1	1.06	1.05	5.7	28.7
28	lignite	Binxin	65.6	4.09	20.66	1.1	0.17	6.2	31.5
29	lignite	Guoleizhuang	66.4	3.78	17.68	1.04	0.94	5.7	26.6

Fig. 2-4 and Fig. 2-5 show the specific comparison results of analysis of different coal elements. With the decrease of the degree of coalification, the content of carbon elements in the injection coal decreases substantially, while the content of oxygen elements and hydrogen elements gradually increases. That is to say, when the blast furnace injection coal develops from anthracite to bituminous coal and lignite, the content of C elements in the injection fuel will decrease, the content of O and H elements will increase, and the content of N and S elements will have little difference. Further comparative analysis of H/C and O/C ratio shows that H/C ratio fluctuates little with the changes of coal types, while O/C ratio fluctuates greatly with the changes of coal types. In particular, the O/C ratio of low-metamorphic coal such as gas coal and lignite coal is much higher than that of high-metamorphic anthracite. The fundamental reason for the above

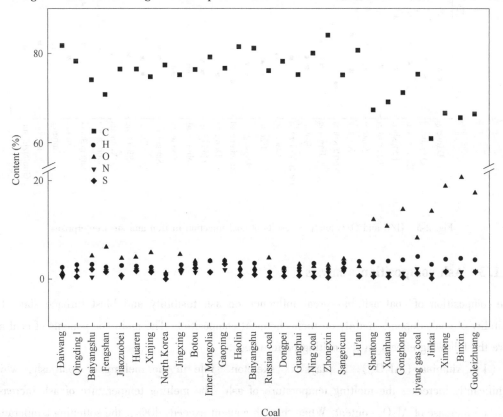

Fig. 2-4 Results of ultimate analysis of coal for injection in iron and steel enterprises

differences is that with the increase of the metamorphism degree of coal, the oxygen-containing functional groups of coal are gradually reduced, the aromatic carbon structure in the carbon structure is gradually increased, and the order degree of the carbon-based structure is increased. Therefore, when a high proportion of bituminous coal and lignite is injected, it is necessary to pay attention to the theoretical combustion temperature reduction caused by the change of C, O and other elements in pulverized coal.

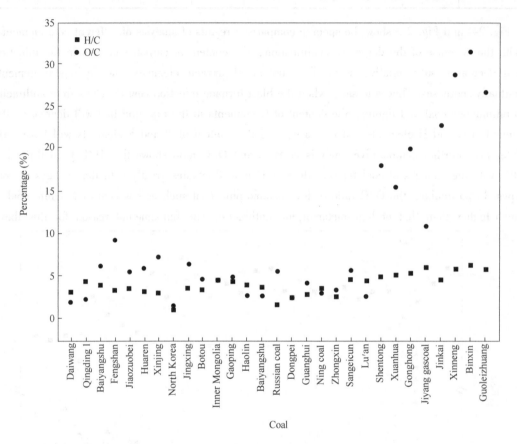

Fig. 2-5 H/C and O/C analysis results of coal injection in Iron and steel enterprises

2.1.3 Ash composition analysis

The composition of coal ash has great influence on ash fusibility and blast furnace slag. The fusibility of coals depends mainly on their chemical composition. The main components of coal ash have the following effects on its melting performance.

(1) Alumina (Al_2O_3). It plays a "skeleton" role in the melting of coal ash, which significantly increases the melting temperature of ash. The melting temperature of ash increases with the increase of Al_2O_3 content. When the ash content exceeds 40%, the softening temperature of coal ash generally exceeds 1,500 ℃.

(2) Silica (SiO_2). In the process of coal ash melting, it has the function of "melting aid",

especially when the content of alkaline components in coal ash is high, the melting aid effect is more obvious. However, the relationship between the content of SiO_2 and ash melting temperature is not obvious. Generally speaking, the ash melting temperature of SiO_2 content more than 40% is about 100℃ higher than that of ash melting temperature of less than 40%. When the content of SiO_2 is in the range of 45%-60%, the melting temperature decreases with the increase of its content.

(3) Iron oxide. In the weakly reducing atmosphere, iron oxide exists in the form of FeO, and the melting temperature of coal ash begins to decrease with the increase of FeO content. When the molar percentage of FeO increases, the melting temperature increases. In the oxidizing atmosphere, iron oxide exists in the form of Fe_2O_3, which always plays a role in raising the melting temperature.

(4) Calcium oxide (CaO). It plays a role in the melting of coal ash, but when its content exceeds a certain limit (CaO in coal ash exceeds 30%), it plays a role in increasing the melting temperature.

(5) Others. Magnesium oxide, sodium oxide and potassium oxide all play a role in the melting of coal ash. The ash in blast furnace coal injection eventually enters into the blast furnace slag. Its influence on the blast furnace slag is that the coal ash contains more SiO_2, Al_2O_3 accounting for 60%-80%, and the rest is a small amount of CaO and MgO, which will generally form acid slag. Therefore, it is necessary to add solvent to make alkaline slag, so as to reach the basicity of blast furnace slag. Therefore, the pulverized coal used in blast furnace coal injection should be selected as far as possible with less ash.

The ash composition analysis of coal injection in blast furnace is shown in Table 2-6. As can be seen from the table, the law of ash composition of different types of coal such as anthracite, bituminous coal and lignite is not obvious, indicating that the coal ash composition has little relationship with the degree of coal metamorphism. For example, in bituminous coal and anthracite, there are coal with higher or lower sulfur content, alkali metal (potassium, sodium) content and ash alkalinity that are usually concerned about in blast furnace smelting. In fact, the composition of pulverized coal ash is closely related to the geological environment of raw coal, and there are significant differences in pulverized coal ash content in different mining areas.

Table 2-6 Analysis of ash composition of coal injection in domestic iron and steel enterprises

(%)

Serial number	Species	Coal	CaO	SiO_2	R_2	Al_2O_3	MgO	Fe_2O_3	SO_3	TiO_2	Na_2O	K_2O
1	anthracite	Haolin	7.27	46.96	0.15	24.84	2.63	7.12	4.60	0.81	0.33	2.95
2	anthracite	Baiyangshu	1.76	49.60	0.04	40.82	0.37	2.24	0.41	1.67	1.06	0.77
3	anthracite	Russian coal	0.99	59.89	0.02	28.39	1.01	5.37	0.29	0.80	0.23	2.62
4	anthracite	Dongpei	21.59	22.47	0.96	11.41	4.47	12.96	20.28	0.64	3.97	0.40
5	anthracite	Guanghui	39.49	18.22	2.17	10.61	2.05	7.86	16.71	0.44	2.97	0.26

Continued Table 2-6

Serial number	Species	Coal	CaO	SiO$_2$	R$_2$	Al$_2$O$_3$	MgO	Fe$_2$O$_3$	SO$_3$	TiO$_2$	Na$_2$O	K$_2$O
6	anthracite	Ning coal	15.62	38.84	0.40	16.04	4.39	7.65	12.07	1.05	2.09	1.39
7	anthracite	Zhongxin	10.19	45.35	0.22	24.78	1.49	5.28	7.05	1.55	1.43	1.95
8	lean coal	Sangeicun	23.82	26.84	0.89	15.15	5.34	10.91	11.61	0.94	3.51	0.45
9	bituminous coal	Shentong	14.19	43.59	0.33	19.89	2.07	8.44	8.17	0.64	1.34	0.95
10	bituminous coal	Xuanhua	12.41	47.72	0.26	19.93	2.18	8.16	5.50	0.66	1.15	1.27
11	bituminous coal	Gonghong	17.80	44.14	0.40	22.43	2.38	4.89	5.19	0.76	0.61	0.73
12	lignite	Jinkai	26.16	30.20	0.87	16.98	7.33	4.80	12.17	0.65	0.43	0.40

2.1.4 Ash melting point

Coal ash is a mixture of multiple components with no fixed melting point and only a melting temperature range. Coal ash is the product of the combustion of minerals in coal at a higher temperature, mainly containing silicate, carbonate, sulfate and sulfide such as silicon, aluminum, iron, calcium, magnesium, potassium and sodium, as well as kaolin and quartz. Most of them are oxidized or decomposed after high temperature burning. The content and properties of these products determine the fusibility of coal ash. At a certain temperature, each component of coal ash will form a eutectic, which can melt the high melting point components of coal ash in the melting state, thus changing the melt composition and melting temperature. As a result, coal ash begins to melt at a lower temperature than the melting point of any single mineral component. Too low ash melting point of pulverized coal is unfavorable to blast furnace injection and even leads to slag formation before tuyere or spray gun. When the ash melts, it prevents oxygen from entering the pulverized coal particles that have not yet burned, resulting in incomplete combustion. In addition, too low ash melting point will accelerate the accumulation and deposition of pulverized coal particles. Therefore, it is generally hoped that the ash melting point of pulverized coal injection is slightly higher, more favorable. The ash melting point of pulverized coal is related to the composition of coal ash. Generally, if the content of coal ash is high in silicon and aluminum, the ash melting point is high; if the content of calcium and iron is high, the ash melting point is low. Because Fe_2O_3 has a low melting point, CaO and SiO_2 will form a low melting point eutectic. The increase of Fe_2O_3 and CaO in ash can reduce the amount of flux added and slag amount, so the use of these coals for injection has more advantages.

The fusibility of coal ash is a physical state which characterizes the deformation, softening and flow of coal ash with the change of heating temperature under certain conditions. When the coal ash sample is heated under specified conditions, with the increase of temperature, the coal ash sample will be from partial melting to total melting, and accompanied by a certain physical state-deformation softening, spheroidal and flowing. The fusibility of coal ash is characterized by the temperature corresponding to the four characteristic physical states.

Fusibility of coal ash is an important quality index of coal. The melting temperature of coal ash can reflect the dynamics of minerals in the blast furnace and can predict the slagging situation. In the above characteristic temperature, softening temperature is widely used, generally according to it to choose the right combustion equipment, according to the type of combustion equipment to choose the raw material coal with the right softening temperature. For example, liquid slagging requires the use of coal at a low melting temperature. Blast furnace is the gasification equipment of liquid slagging, the temperature is high, the range of adaptation to the temperature of coal ash melting is wide. The melting temperature of coal ash is low, and the remaining part of coarse-grained pulverized coal after combustion is easy to be wrapped by molten ash, which is unfavorable to improving the combustion rate of pulverized coal. Blast furnace slag should have good fluidity, if the temperature of complete flow of coal ash is too high, it is unfavorable to desulfurization.

Ash fusibility was determined according to GB/T 219—2008. Table 2-7 shows the classification of coal ash softening temperature.

Table 2-7 Classification of coal ash softening temperature

Serial number	Level name	Code	Grading range ST(℃)
1	low softening temperature ash	LST	≤1,100
2	lower softening temperature ash	RLST	>1,100-1,250
3	medium softening temperature ash	MST	>1,250-1,350
4	medium high softening temperature ash	RHST	>1,350-1,500
5	high softening temperature ash	HST	>1,500

Table 2-8 shows the four characteristic temperatures of ash fusibility of typical coal in iron and steel enterprises, among which the softening temperatures of coal are shown in Fig. 2-6. It can be found that the softening temperatures of different coal types in iron and steel enterprises are significantly different, even if the softening temperatures of the same type of coal are significantly different, and the softening temperature classification of coal ash has a large span, which is distributed between low softening temperature ash and high softening temperature ash. For example, the softening temperature of Baiyangshu coal in anthracite is greater than 1,510℃, while that of Zhongxin coal in anthracite is only 1,135℃. With the increase of the degree of coalification, the change of ash melting point does not show an orderly law, that is, compared with the ash melting point of anthracite, lignite and bituminous coal have no obvious characteristics. Therefore, when using lignite and bituminous coal, iron and steel enterprises should measure and evaluate the ash melting characteristics of coal separately, so as to avoid the problems of slagging and erosion of blowpipe caused by low ash melting point.

Table 2-8 Characteristic temperature of ash melting characteristics of different coal types in iron and steel enterprises (℃)

Serial number	Species	Coal	Deformation Temperature (DT)	Soft melting Temperature (ST)	Hemispheric Temperature (HT)	Flow Temperature (FT)
1	anthracite	Dariwang	1,305	1,315	1,395	1,405
2	anthracite	Qingding 1	1,450	1,510	>1,510	>1,510
3	anthracite	Baiyangshu	1,440	>1,510	>1,510	>1,510
4	anthracite	Fengshan	1,300	1,310	1,345	1,385
5	anthracite	Jiaozuobei	1,290	1,420	1,495	1,510
6	anthracite	Huaren	1,395	1,410	1,420	1,430
7	anthracite	Xinjing	1,220	1,245	1,320	1,350
8	anthracite	North Korea	1,255	1,390	1,465	1,485
9	anthracite	Jingxing	1,295	1,405	1,455	1,480
10	anthracite	Botou	>1,510	>1,510	>1,510	>1,510
11	anthracite	Inner Mongolia	1,330	1,350	1,375	1,405
12	anthracite	Gaoping	1,365	1,375	1,385	1,400
13	anthracite	Haolin	1,190	1,210	1,230	1,240
14	anthracite	Baiyangshu	1,470	>1,470	>1,470	>1,470
15	anthracite	Russian coal	1,360	1,420	1,450	1,470
16	anthracite	Dongpei	1,130	1,194	1,242	1,254
17	anthracite	Gaunghui	1,133	1,151	1,162	1,163
18	anthracite	Ning coal	1,158	1,192	1,216	1,250
19	anthracite	Zhongxin	1,133	1,135	1,149	1,168
20	lean coal	Sangeicun	1,340	1,405	1,500	>1,510
21	lean coal	Lu'an	1,330	1,365	1,445	1,475
22	bituminous coal	Shentong	1,100	1,120	1,130	1,170
23	bituminous coal	Xuanhua	1,150	1,160	1,190	1,200
24	bituminous coal	Gonghong	1,200	1,240	1,280	1,300
25	gas coal	Jiyang gas coal	>1,510	>1,510	>1,510	>1,510
26	lignite	Jinkai	1,230	1,250	1,270	1,270
27	lignite	Xinneng	1,215	1,220	1,225	1,230
28	lignite	Binxin	1,165	1,185	1,200	1,205
29	lignite	Guoleizhuang	1,200	1,220	1,230	1,250

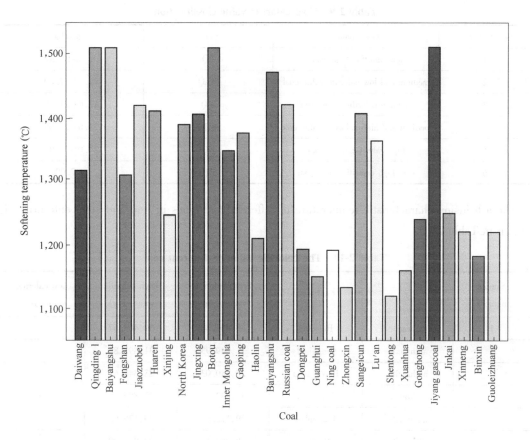

Fig. 2-6 Softening temperature distribution of different coal species

2.1.5 Calorific value

The calorific value of coal refers to the heat released by the combustion of unit coal (kg or g). The calorific value of general coal is measured according to the national standard GB 213—1987, expressed by J/g, kJ/kg and MJ/kg.

(1) Bomb caloric value Q_b. The calorific value of unit mass coal combustion in oxygen bomb filled with excess oxygen was directly measured by oxygen bomb calorimeter in the laboratory. The final products were carbon dioxide at 25℃, excess oxygen, nitrogen, nitric acid, sulfuric acid, liquid water, and heat released from solid ash.

(2) Overall heating value Q_{gr}. The metallurgical industry stipulates that the heat released when the combustion product cools to the water vapor condensed into 0℃ water after the fuel is completely burned.

(3) Low calorific value Q_{net}. The heat released when the water vapor in the combustion product cools to 20℃ after the fuel is completely burned.

The classification of coal calorific value in China is divided according to the size of high calorific value, as shown in Table 2-9.

Table 2-9 Coal calorific value classification

Serial number	Level name	Code	Q(MJ/kg)
1	low calorific value coal	LQ	8.50-12.50
2	medium and low calorific value coal	MLQ	12.51-17.00
3	medium calorific value coal	MQ	17.01-21.00
4	medium and high calorific value coal	MHQ	21.01-24.00
5	high calorific value coal	HQ	24.01-27.00
6	very high calorific value coal	SHQ	>27.00

The ash fusion characteristic temperature of different coal types in iron and steel enterprises is shown in Table 2-10.

Table 2-10 The calorific value of different coal

Sample number	Class	Coal dust	Bomb calorific value(J/g)	High calorific value(J/g)	Low calorific value(J/g)
1	anthracite	Daiwang	30,457.40	30,408.67	—
2	anthracite	Qingdin 1	31,949.93	31,898.81	—
3	anthracite	Baiyangshu	28,380.03	28,340.61	—
4	anthracite	Fengshan	27,932.48	27,887.79	—
5	anthracite	Jiaozuobei	29,172.62	29,125.95	—
6	anthracite	Huaren	28,860.15	28,813.98	—
7	anthracite	Xinjing	27,941.35	27,896.64	—
8	anthracite	Korea	25,541.47	25,500.61	—
9	anthracite	Jingxing	28,606.34	28,560.57	—
10	anthracite	Botou	28,857.9	28,811.73	—
11	anthracite	Inner Mongolia	31,015.44	30,961.35	—
12	anthracite	Gaoping	28,977.68	28,928.57	—
13	anthracite	Haolin	31,427.26	31,376.97	30,734.25
14	anthracite	Baiyangshu	31,233.76	31,183.78	30,576.08
15	anthracite	Russian coal	27,303.36	27,259.69	27,028.97
16	anthracite	Dongpei	—	29,446.12	—
17	anthracite	Guanghui	—	29,197.57	—
18	anthracite	Ning coal	—	30,544.97	—
19	anthracite	Zhongxin	—	31,634.06	—
20	non-gaseous coal	Sangeicun	29,507.14	29,459.93	—
21	non-gaseous coal	Lu'an	31,627.25	31,576.65	—
22	bituminous coal	Shentong	26,087.43	26,045.69	25,367.95

2.1 Metallurgical properties of bituminous coal and lignite · 23 ·

Continued Table 2-10

Sample number	Class	Coal dust	Bomb calorific value(J/g)	High calorific value(J/g)	Low calorific value(J/g)
23	bituminous coal	Yihua	26,656.75	26,614.41	25,907.83
24	bituminous coal	Gonghong	27,830.77	27,786.24	27,024.04
25	gas coal	Luoyang gas coal	30,667.47	30,718.41	—
26	brown coal	Jingkai	23,089.93	23,062.23	22,493.67
27	brown coal	Xinneng	25,867.58	25,827.19	—
28	brown coal	Binxin	25,967.32	25,037.24	—
29	brown coal	Guoleizhuang	23,969.80	23,941.04	—

The high calorific value of different coal types are shown in Fig. 2-7. It can be seen from the test results of the high calorific value of the injected coal in the iron and steel enterprises that the calorific value of the injected coal in the blast furnace is basically greater than 24MJ/g, which belongs to the high calorific value coal, and some coals are even ultra-high calorific value coal.

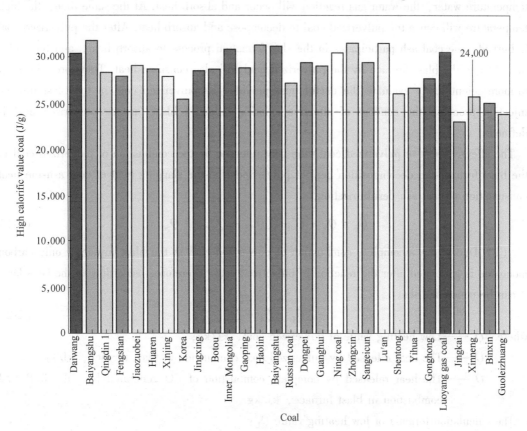

Fig. 2-7 High calorific value of different coals

In actual production, once the pulverized coal particles injected into the blast furnace enter the hot air of the direct blowing tube, they are exposed to an oxidizing atmosphere, and the pyrolysis reaction and combustion reaction occur simultaneously under rapid heating conditions. The decomposition of pulverized coal is endothermic, and the higher the volatile matter of pulverized coal, the more endothermic it is.

The combustion of pulverized coal injected into the blast furnace before the tuyere is different from that of coal in the atmosphere or in the boiler. In the tuyere of the blast furnace, carbon is more than oxygen, and the temperature is higher. After the pulverized coal is burned, it can only be oxidized to CO, H_2, N_2, and cannot be burned in the atmosphere to produce CO_2, H_2O, N_2. That is to say, the heat released by carbon combustion in pulverized coal is much lower than that released by combustion in the atmosphere. H_2 can not be burned in the tuyere zone, and does not release heat. On the contrary, the organic matter in the coal is decomposed and the heat is absorbed. The decomposition heat of pulverized coal can be defined as the heat absorbed by C, H, CO and N_2 per kilogram of pulverized coal under high temperature inert gas conditions. The decomposition heat of bituminous coal is greater than that of anthracite. In addition, after the pulverized coal of the blast furnace enters the direct blowing tube, due to the volatilization of high temperature water, the water gas reaction will occur and absorb heat. At the same time, the high temperature will cause the pulverized coal to decompose and absorb heat. After the pulverized coal is burned, the coal ash participates in the slag formation process to absorb heat, and the sulfur brought into the blast furnace by the pulverized coal will also consume heat. Therefore, in order to more scientifically evaluate the useful heat provided by pulverized coal in the blast furnace injection process, the concept of effective calorific value of blast furnace pulverized coal is defined.

The effective heat of pulverized coal = the heat released by the combustion of pulverized coal in the blast furnace-the decomposition heat of pulverized coal-the slagging heat of coal ash-the heat consumption of coal ash desulfurization.

$$Q_e = Q_1 - Q'_d - (Q_{AS} + Q_{AP}) - Q_S \tag{2-1}$$

(1) Due to the incomplete combustion of pulverized coal in the blast furnace, only carbon monoxide is generated after the reaction of pulverized coal. Therefore, according to the Geis law, it can be concluded that:

$$Q_L = Q_1 + Q_2 \tag{2-2}$$

where Q_L ——Low heating value of pulverized coal, kJ/kg;

Q_1 ——The heat released by pulverized coal combustion in blast furnace, kJ/kg;

Q_2 ——The heat released by complete combustion of CO generated by pulverized coal combustion in blast furnace, kJ/kg.

The calculation formula of low heating value Q_L:

$$Q_L = Q_H - 25.12(w(H_2O) + 9w(H)) \tag{2-3}$$

where Q_H ——High heating value of pulverized coal, kJ/kg;

$w(H_2O)$ ——The mass fraction of crystal water in pulverized coal, %;

$w(H)$——The mass fraction of hydrogen in pulverized coal, %.

The high calorific value of pulverized coal in the tuyere is 28,053kJ/kg measured by the pulverized coal calorific value measuring instrument. The ultimate analysis of coal in the tuyere under the condition of air drying base is shown in Table 2-11.

Table 2-11 Ultimate analysis of blast tuyere Coal (%)

Coal	M_{ad}	C_{ad}	H_{ad}	N_{ad}	O_{ad}	$S_{t,ad}$
Tuyere coal	1.50	78.09	3.75	1.04	5.59	0.47

The crystal water in the pulverized coal injected into the blast furnace exists in the form of ferrous sulfate. The crystal water content of bituminous coal is generally about 0.1%. The ratio of mixed coal in the blast furnace in industrial production is bituminous coal : anthracite = 43 : 57. Therefore, the crystal water content in the tuyere coal is 0.05%. According to Eq. (2-3), the low heating value of pulverized coal can be calculated. According to Eq. (2-2), the heat released by pulverized coal combustion in blast furnace is calculated. The high heating value of blast furnace mixed coal is 28,053kJ/kg, and the low heating value is 27,249kJ/kg, The reaction of CO combustion:

$$CO + \frac{1}{2}O_2 = CO_2 + \Delta H_2, \quad \Delta H_2 = -283.4 \text{kJ/mol} \qquad (2-4)$$

So $\qquad Q_2 = 18,421\text{kJ/kg}, \quad Q_1 = 8,828\text{kJ/kg}.$

(2) The heat brought out of the blast furnace by coal ash is divided into two parts: one part is the sensible heat of ash, and the other part is the phase change melting and melting of ash from the temperature of pulverized coal carrier gas to the temperature of slag. The constant pressure heat capacity of ash is expressed by Eq. (2-5):

$$c_p = A + Bt \qquad (2-5)$$

where A, B——Constant pressure specific heat capacity coefficient of different substances;

$\qquad t$——Kelvin temperature, K.

The constant pressure specific heat capacity of unit mass coal ash is:

$$c_{p_A} = n(SiO_2)c_{p_{SiO_2}} + n(Al_2O_3)c_{p_{Al_2O_3}} + n(CaO)c_{p_{CaO}} + n(MgO)c_{p_{MgO}} \qquad (2-6)$$

The constant pressure specific heat capacity coefficients of different substances in ash are shown in Table 2-12.

Table 2-12 Constant pressure specific heat capacity coefficient of main substances in ash

Ash composition	A(J/K/mol)	B(J/K/mol)
SiO_2	114.8	0.01280
Al_2O_3	48.83	0.00452
CaO	42.59	0.00728
MgO	46.94	0.03431

c_{p_A} : 1,053.56+0.135t; Q_{AS} : the sensible heat of coal ash.

$$Q_{AS} = \eta_A \int_{t_0}^{t_1} c_{p_A} dt \tag{2-7}$$

where t_0——Carrier gas temperature, K;

t_1——Slag temperature, K;

η_A——The mass fraction of ash in pulverized coal, %.

The carrier gas temperature is 25℃, the hot metal temperature is 1,500℃, The mass fraction of ash in pulverized coal η_A is 7.86%, The slag is 50℃ higher than hot metal, Substituting c_{p_A} into Eq. (2-7), $Q_{AS} = 161.41$ kJ/kg.

Phase change enthalpy and melting enthalpy per unit mass of pulverized coal $Q_{AP} = \eta_A Q_{AU}$, The temperature in the blast furnace is not enough to melt calcium oxide and magnesium oxide, so the melting enthalpy of the three is not considered, so Q_{AU} is:

$$Q_{AU} = (251.7 + 21.7) w(SiO_2) + 844.7 w(Al_2O_3) \tag{2-8}$$

The calculated Q_{AU} is 347.5 kJ/kg. The phase change enthalpy and melting enthalpy Q_{AP} of ash in unit mass of pulverized coal are 27.31 kJ/kg.

(3) Coal ash desulfurization heat consumption. The sulfur in blast furnace comes from raw materials. The sulfur in pulverized coal mainly exists in the form of organic sulfur and inorganic sulfur. Inorganic sulfur mainly comes from various sulfur-containing compounds in minerals. There are mainly sulfide sulfur and a small amount of sulfate sulfur, and occasionally elemental sulfur. The sulfur in pulverized coal ash will absorb heat by desulfurization reaction in blast furnace. The heat consumption of desulfurization is mainly composed of sulfide decomposition into S and S into slag to become CaS.

$$FeS \Longrightarrow Fe + S \qquad \Delta H = 2,990 \text{kJ/kgS} \tag{2-9}$$

$$FeS_2 \Longrightarrow Fe + 2S \qquad \Delta H = 2,780 \text{kJ/kgS} \tag{2-10}$$

$$CaO + S \Longrightarrow CaS + 0.5O_2 \qquad \Delta H = 5,410 \text{kJ/kgS} \tag{2-11}$$

The desulfurization heat consumption of coal ash $Q_S = 8,190 \times 8,400 \times S_{MS}$, S_{MS} is the mass of sulfur entering the slag in tons of iron pulverized coal. According to the sulfur balance in the blast furnace, the sulfur brought into the furnace charge = the sulfur entering the slag + the sulfur escaping from the furnace with the gas + the sulfur contained in the pig iron. Calculated according to the amount of sulfur into the furnace with 10% of the gas.

$$0.9(O_S + C_S + M_S) - I(S) = S_S \tag{2-12}$$

where O_S——The amount of sulfur brought into tons of iron ore;

C_S——The amount of sulfur brought in by coke per ton of iron;

M_S——Tons of iron pulverized coal into the amount of sulfur;

$I(S)$——The amount of sulfur dissolved in molten iron;

S_S——The amount of sulfur taken away by tons of iron slag.

Therefore, the amount of sulfur brought into the blast furnace by pulverized coal and finally into the slag is:

$$S_{MS} = \frac{M_S}{O_S + C_S + M_S} \times S_S \tag{2-13}$$

According to the production report of blast furnace, coke ratio is 322kg/tHM, coal ratio is 148kg/tHM, coke ratio is 32kg/tHM, sulfur content in coke is 0.79%, sulfur content in pulverized coal is 0.46%, sulfur content in molten iron is 0.03%. The ratio of the amount of sulfur brought in by all-day sinter, oxygen ball, titanium ore and Australian ore to the output of all-day blast furnace is the amount of sulfur brought in by tons of iron raw materials. The sulfur content of blast furnace raw materials is shown in Table 2-13.

Table 2-13 Sulfur content of raw materials into the furnace (%)

Raw material	Sintering ore	Oxygen ball	Titanium ore	Australian mine
S contents	0.025	0.0032	0.05	0.013

The amount of sulfur O_S brought by tons of iron raw materials is 0.322kg, the amount of sulfur brought by tons of iron coke is 2.7kg, the amount of sulfur brought by tons of iron pulverized coal is 0.68kg, and the total amount of sulfur brought by tons of iron is 3.79kg. The amount of sulfur in per ton iron is 0.3kg. S_S is 3.12kg. Bring the data into the formula, S_{MS} is 0.56kg, and the heat consumption of desulfurization per ton of iron is (4,582-4,693) kJ/kg. Because the coal ratio is 148kg/tHM, the heat consumption of desulfurization per unit mass of pulverized coal is Q_S = (30.96-31.71) kJ/kg.

(4) Pulverized coal decomposition heat. Through the new calculation method of pulverized coal decomposition heat proposed in this study, the heat absorption of pulverized coal decomposition in front of the tuyere is calculated. Pulverized coal will be endothermically decomposed into C, CO, H_2, N_2, etc. after entering the direct blow pipe. According to Geis law, the thermal effect of chemical reaction is only related to its initial and final states. The decomposition heat of pulverized coal is:

$$Q_d = Q_b - Q_t \qquad (2\text{-}14)$$

where Q_d ——The decomposition heat of pulverized coal, kJ/kg;

Q_b ——Pulverized coal cartridge heating value, kJ/kg;

Q_t ——The total heat released by the reaction of elements in pulverized coal with oxygen after decomposition and cooling to room temperature, kJ/kg.

According to the ultimate analysis and ash content of tuyere coal, the molar amounts of C, CO, H_2, and N_2 after the decomposition of 1kg pulverized coal dry basis can be calculated respectively. The composition analysis of the coal ash in the tuyere is shown in Table 2-14.

Table 2-14 Ash composition analysis of blast furnace tuyere coal (%)

Ash composition	SiO_2	Al_2O_3	CaO	Fe_2O_3	SO_3	MgO	Na_2O	TiO_2	P_2O_5
Percentage	43.53	27.05	8.35	7.76	4.19	2.36	2.77	1.64	0.86

According to the ultimate analysis of the tuyere coal in Table 2-14, the mass percentage of oxygen is 5.59%. Part of the oxygen in the pulverized coal reacts to form CO, and part of it exists in the ash. Due to the good chemical stability of the ash, only the oxygen on the carbon chain decomposes to form CO. The ash content of tuyere coal is 9.79%. The weighted sum of different components in ash is used to calculate the oxygen content:

$$A_O = 43.53 \times 32/60 + 27.05 \times 48/102 + 8.35 \times 16/56 + \cdots + 0.86 \times 80/142 \tag{2-15}$$

$A_O = 45.97$, 1kg of pulverized coal into the coal ash in the amount of oxygen is 0.045kg, the amount of oxygen generated by CO is 0.0109kg, the generated CO is 0.68125mol, the amount of carbon formed after the decomposition of pulverized coal is 64.39mol, and the amount of hydrogen is 18.75mol. According to the weighted calculation of enthalpy of different components, it can be concluded that:

$$Q_t = \Delta H_c n_c + \Delta H_{CO} n_{CO} + \Delta H_{H_2} n_{H_2} \tag{2-16}$$

The calculated Q_t is 31,053.2kJ/kg, the elastic heating value Q_b of pulverized coal is 28,059.37kJ/kg, then Q_d is 2,993.83kJ/kg, and the effective heat Q_e of pulverized coal is (5,613.74–5,614.49)kJ/kg.

The data of heat consumption such as combustion heat release and decomposition desulfurization in the process of coal injection are shown in Table 2-15.

Table 2-15 Heat data of pulverized coal in blast furnace   (kJ/kg)

Q_e	Q_1	Q_d	Q_{AS}	Q_{AP}	Q_S
5,613.74-5,614.49	8,828	2,993.83	161.41	27.31	30.96-31.71

From Table 2-15, it can be seen that the heat consumption of ash slagging and decomposition accounts for 1.82% and 33.91% of the heat released by pulverized coal combustion, respectively. The effective heat of pulverized coal accounts for 63.59% of the heat released by pulverized coal combustion. Because the pulverized coal in the blast furnace will certainly undergo a decomposition process, this part of the heat loss can not be avoided, and the size of the ash slagging heat consumption is related to the ash content of the pulverized coal into the furnace. The ash content in the pulverized coal is reduced, and the effective calorific value of the pulverized coal is increased. Therefore, the ash content in the pulverized coal injected into the blast furnace should be controlled.

2.1.6 Grindability

The grindability index of coal marks the difficulty of coal crushing.

Pulverized coal is usually used in blast furnace to facilitate combustion. Therefore, it must be ground. Grindability of coal refers to the degree of difficulty of grinding coal, which is mainly related to the degree of metamorphism of coal. Generally speaking, the grindability index of coking coal and fat coal is higher, that is, it is easy to grind; the grindability index of anthracite and lignite is low, which is not easy to grind. In addition, the grindability also decreases with the increase of moisture and ash content of coal, that is, the higher the moisture and ash content of the same coal, the lower the grindability index. In industry, the mill is designed according to the grindability, the yield and energy consumption of the mill are estimated, or the coal type and coal source suitable for a specific type of mill are selected according to the grindability of coal.

The grindability index of coal for blast furnace injection should be between 60 and 90, and the coal below 50 is hard and difficult to grind. Although bituminous coal above 90 is easy to grind, it is often strongly caking coal, which may cause difficulties in coal grinding and coal transportation. There are two ways to express the grindability coefficient, namely the Hardgrove coefficient and the вги coefficient. The former is mainly used in Western European countries and the United States, while the latter is used in Eastern European countries. At present, the вги coefficient has been replaced by the Hardgrove coefficient because of its inconvenient operation and poor reproducibility. At present, the grindability index-HGI of coal is determined by Hardgrove method in China, which is defined as:

$$H = 13 + 6.93W$$

In the formula, W is the pulverized coal passing through 200 mesh (75μm) after the coal is ground by the Hardgrove grindability tester.

The coefficient of the Hardgrove method is based on the Pennsylvania coal in the United States as the standard, that is, $H=100$ is a coal with good grindability (i.e., softer coal, which is similar to China's peak coal). Compared with other coals, the smaller the H, the more difficult the coal is to grind. In addition, the research shows that the coalification degree is the main factor affecting the grindability coefficient of coal. The coal with high and low coalification degree has poor grindability, and the coal with medium coalification degree has good grindability. The grindability index was determined according to the national standard GB/T 2565—2014. The grindability index of blast furnace injection coal commonly used in iron and steel enterprises is shown in Table 2-16 and Fig. 2-8. The analysis results on the data are basically consistent with the usual law, that is, the grindability of anthracite is low, followed by the grindability of lignite, and the grindability of bituminous coal and lean coal is high. However, there are also abnormal grindability of individual coal types, such as Haolin anthracite and Xinneng lignite. Therefore, to solve the problem of poor grindability of lignite injection in blast furnace, we can break through from two aspects: on the one hand, by optimizing the high grindability of lignite, on the other hand, by optimizing the matching of coal types.

Table 2-16 Grindability index of common injection coal in blast furnace

Sample number	Type	Coal	HGI
1	anthracite	Daiwang	33.8
2	anthracite	Qingdin 1	78.8
3	anthracite	Baiyangshu	75.4
4	anthracite	Fengshan	71.9
5	anthracite	Jiaozuobei	47.7

Continued Table 2-16

Sample number	Type	Coal	HGI
6	anthracite	Huaren	44.2
7	anthracite	Xinjing	61.5
8	anthracite	Korea	75.4
9	anthracite	Jingxing	68.5
10	anthracite	Botou	78.8
11	anthracite	Inner Mongolia	82.3
12	anthracite	Gaoping	89.3
13	anthracite	Haolin	103.09
14	anthracite	Baiyangshu	92.7
15	anthracite	Russian coal	75.37
16	anthracite	Dongpei	51.1
17	anthracite	Guanghui	37.3
18	anthracite	Ning coal	51.6
19	anthracite	Zhongxin	44.2
20	lean coal	Sangeicun	106.6
21	lean coal	Lu'an	106.6
22	bituminous coal	Shentong	99.63
23	bituminous coal	Yihua	99.63
24	bituminous coal	Gonghong	64.98
25	gas coal	Luoyang gas coal	71.9
26	lignite coal	Jingkai	83.69
27	lignite coal	Xinneng	117
28	lignite coal	Binxin	89.3
29	lignite coal	Guoleizhuang	82.3

2.1.7 Adhesion

When the crushed coal isolated air is gradually heated to 200-500℃, a part of the gas will be separated out and a viscous colloid will be formed. When it is further heated to 500℃, the viscous colloid continues to decompose, part of it is decomposed into gas, and the rest is

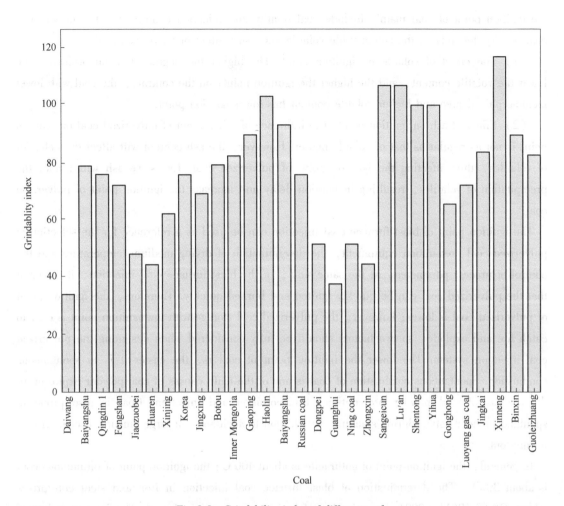

Fig. 2-8 Grindability index of different coals

gradually solidified. The carbon particles are combined to form coke blocks. The degree of firmness of this combination is called cohesiveness. The cohesion index is a key index to judge the cohesion and coking properties of coal, which is used to evaluate the cohesion ability of bituminous coal in the heating process, expressed as $G_{R.I.}$. The cohesion index is used to evaluate the cohesion ability of bituminous coal in the heating process. The experimental method for measuring the cohesion index of blast furnace injection coal follows the GB/T 5447—85 bituminous coal cohesion index measurement method. Coal for blast furnace injection requires no cohesiveness, that is, the cohesion index is 0.

2.1.8 Ignition point

The ignition point of coal refers to the temperature at which coal is heated to the beginning of combustion under the condition of coexistence of oxidant (air) and coal. In scientific language, the minimum ignition temperature at which coal releases enough volatiles to form a combustible mixture with the surrounding atmosphere is called the ignition point of coal. The factors affecting the ignition point of coal mainly include coal quality and injection conditions. The influence of coal quality depends on the combustible volatile and ash content of pulverized coal.

(1) The effect of volatile on ignition point. The higher the degree of metamorphism, the lower the volatile content, and the higher the ignition point; on the contrary, the coal with lower metamorphic degree and higher volatile content has lower ignition point.

(2) Effect of ash on ignition point. The influence of ash content of pulverized coal on ignition point is not as regular as that of volatile matter. However, the ash content will affect the behavior of volatiles, thus affecting the ignition point of pulverized coal. Excessive ash will hinder the precipitation of volatiles, resulting in ignition delay and increase the ignition point of pulverized coal.

The ignition point of blast furnace coal injection can be used as a reference for the selection of pulverized coal preparation equipment, the determination of drying medium temperature and the control of process parameters. At the same time, in the blast furnace coal injection, it is hoped that the pulverized coal can be quickly ignited and burned quickly. Therefore, the ignition point of pulverized coal is lower; however, the pulverized coal storage with low ignition point is easy to catch fire and explode, so two factors should be fully considered when designing the pulverized coal injection system. The lower the ignition point of coal is, the easier it is to spontaneous combustion. Spontaneous combustion of coal is one of the main causes of coal powder explosion in the process of coal powder preparation, transportation and injection. Coal is also prone to spontaneous combustion during stacking. In addition to accidents, a large amount of coal will be burned out.

In general, the ignition point of anthracite is about 400℃; the ignition point of bituminous coal is about 300℃. The determination of blast furnace coal injection in iron and steel enterprises adopts GB/T 18511—2001 determination of ignition temperature of coal. The ignition points of coal injection commonly used in iron and steel enterprises are shown in Table 2-17.

Table 2-17 The ignition point of coal injection commonly used in iron and steel enterprises

Sample number	Type	Coal	Ignition point (℃)
1	anthracite	Daiwang	359.5
2	anthracite	Qingdin 1	344.5
3	anthracite	Baiyangshu	356.5
4	anthracite	Fengshan	341.3
5	anthracite	Jiaozuobei	357.6
6	anthracite	Huaren	340.8
7	anthracite	Xinjing	355.7
8	anthracite	Korea	399.8
9	anthracite	Jingxing	354.3
10	anthracite	Botou	351.9

2.1 Metallurgical properties of bituminous coal and lignite

Continued Table 2-17

Sample number	Type	Coal	Ignition point (℃)
11	anthracite	Inner Mongolia	338.5
12	anthracite	Gaoping	351.8
13	anthracite	Haolin	410.38
14	anthracite	Baiyangshu	399.75
15	anthracite	Russian coal	412.75
16	anthracite	Dongpei	401
17	anthracite	Guanghui	401
18	anthracite	Ning coal	427
19	anthracite	Zhongxin	373
20	lean coal	Sangeicun	323.2
21	lean coal	Lu'an	317.2
22	bituminous coal	Shentong	318.55
23	bituminous coal	Yihua	325.83
24	bituminous coal	Gonghong	322.28
25	gas coal	Juoyang gas coal	291.9
26	lignite coal	Jingkai	312.73
27	lignite coal	Xinneng	258.5
28	lignite coal	Binxin	256.1
29	lignite coal	Guoleizhuang	268.8

From Fig. 2-9, it can be seen that with the increase of the metamorphic degree of pulverized coal, the ignition point of blast furnace injection coal gradually increases, and the ignition point of lignite is the lowest, which is as low as 256.1℃. The ignition point temperature is much lower than the mill inlet temperature, so the lignite injection is prone to the explosion of the pulverizing system. In the production process, lignite injection needs to focus on system air leakage, local high temperature and other issues, while using lignite and other pulverized coal mixed injection to reduce its ignition point.

2.1.9 Explosibility

There are many ways to measure the explosiveness of pulverized coal. In China, a long tube test device is mainly used to measure the flame return length of pulverized coal combustion to determine whether the pulverized coal is explosive and its explosive strength. The device is shown in Fig. 2-10. During the test, 1g -200 mesh (-75μm) pulverized coal samples were weighed and mixed with sodium nitrite at a ratio of 1 : 0.75. After mixing, 1g was taken and sprayed onto a fire source at 1,050℃ in a glass tube. The length of the flame returned to determine its explosiveness. It is generally believed that non-explosive coals, such as anthracite, are only

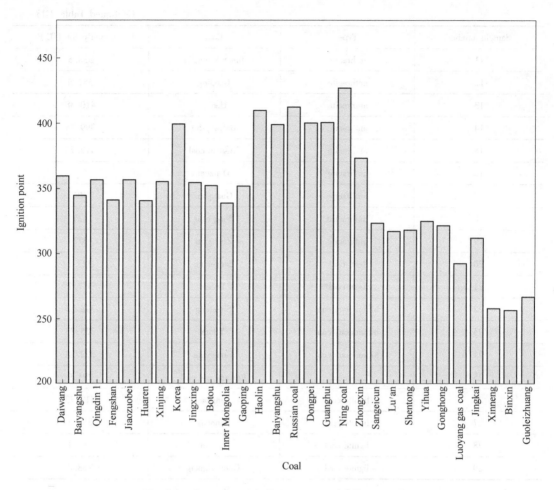

Fig. 2-9 Comparison of ignition points of different coals

considered to have rare or non-Mars at the source of the fire. If the flame is generated and returned to the injection end, the flame length is less than 400mm for flammable and explosive coal; if the return flame is greater than 400mm, it is a strong explosive coal, such as bituminous coal and lignite. It is generally believed that ash-free (combustible) volatile matter less than 10% is non-explosive coal; more than 10% is explosive coal; more than 25% is strong explosive coal. The explosiveness of pulverized coal is also related to its particle size. The finer the pulverized coal is, the easier it is to explode. However, under the same volatile matter and particle size, the explosiveness will be different due to the different specific surface area.

Safety must be paid attention to in the design and operation of pulverized coal injection in blast furnace to prevent pulverized coal explosion. The pulverized coal explosion must have the following three conditions: the pulverized coal is in a dispersed suspension state, and the concentration is within the explosion range; enough oxygen to support combustion; an ignition source with sufficient energy. The explosiveness of coal commonly used in iron and steel enterprises is shown in Table 2-18.

2.1 Metallurgical properties of bituminous coal and lignite

Fig. 2-10 Long tube pulverized coal explosive detection equipment

Table 2-18 Explosive situation of common coal in iron and steel enterprises

Sample number	Type	Coal	Flame length (mm)
1	anthracite	Daiwang	5
2	anthracite	Qingdin 1	5
3	anthracite	Baiyangshu	5
4	anthracite	Fengshan	6
5	anthracite	Jiaozuobei	5
6	anthracite	Huaren	5
7	anthracite	Xinjing	5
8	anthracite	Korea	5
9	anthracite	Jingxing	5
10	anthracite	Botou	5
11	anthracite	Inner Mongolia	5
12	anthracite	Gaoping	5
13	anthracite	Haolin	0
14	anthracite	Baiyangshu	0
15	anthracite	Russian coal	15
16	anthracite	Dongpei	0
17	anthracite	Guanghui	0
18	anthracite	Ning coal	0
19	anthracite	Zhongxin	0
20	lean coal	Sangeicun	5
21	lean coal	Lu'an	5

Continued Table 2-18

Sample number	Type	Coal	Flame length (mm)
22	bituminous coal	Shentong	566.75
23	bituminous coal	Yihua	665.25
24	bituminous coal	Gonghong	720.25
25	gas coal	Luoyang gas coal	8
26	lignite coal	Jingkai	746.25
27	lignite coal	Xinneng	>800
28	lignite coal	Binxin	723
29	lignite coal	Guoleizhuang	673

Through the data of Table 2-18, it can be seen that with the increase of the metamorphic degree of pulverized coal, the explosiveness of blast furnace coal injection is gradually weakened. Lignite and bituminous coal have strong explosiveness, anthracite and lean coal have no explosiveness. Therefore, blast furnace must be explosion-proof when using bituminous coal and lignite to avoid safety accidents.

2.1.10 Flowability and injectability

The newly ground coal powder can absorb gas (such as air) and form a gas film on the surface of coal particles, which makes the frictional resistance between coal particles smaller; in addition, coal particles are all charged bodies, and they are all homogeneously charged. Homogeneous charges have mutual repulsion, so coal particles have good fluidity, which is manifested by a small natural slope angle, and can flow with the carrier in a carrier at a certain speed, which is the basis for conveying coal powder. However, with the extension of coal powder storage time, the gas film on the surface of coal particles becomes thinner, the static electricity gradually disappears, and the fluidity of coal powder will gradually deteriorate. Blast furnace pulverized coal injection needs pneumatic conveying, so coal with good fluidity is required. From this point of view, the storage time of pulverized coal should not be too long, such as Ansteel stipulates that it should be less than 8h.

2.1.10.1 Experimental equipment and methods

The experimental equipment adopts the BT-1000 powder comprehensive characteristic tester jointly developed by Dandong Baite Instrument Co., Ltd. and Tsinghua University Powder Technology Development Department, as shown in Fig. 2-11.

The research on the fluidity of pulverized coal includes the determination of fluidity and jetness. In the experiment, each group measured 3 times, and took the average value. The particle size of pulverized coal in this test is measured according to -200 mesh ($-75\mu m$) $70\%\pm5\%$, which is designed to simulate the actual blast furnace injection particle size.

2.1.10.2 Fluidity

The flow characteristics of pulverized coal include four factors: natural slope angle, compressibility, plate scoop angle and uniformity. Among them, the compressibility is obtained through the loose bulk density and the tapped bulk density, and the others are directly measured by experiments.

(1) Loose bulk density. The test adopts a cylindrical cup-shaped container with a volume of $100cm^3$, coal powder slowly flows into the container from a distance of 100mm from the upper edge of the container, and scrape the upper surface of the container once with a thin iron sheet when the container is full level, weigh

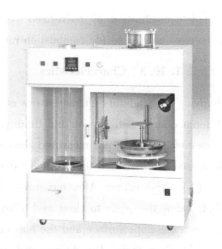

Fig. 2-11 BT-1000 Type Powder Comprehensive Characteristic Tester

the weight of the loaded coal powder, and then divide it by the volume of the container, which is the loose packing density, which can be regarded as the bulk specific gravity.

(2) Vibrated bulk density. Put a container cover of $100cm^3$ on the container after measuring the loose bulk density, after the container cover is filled with coal powder, put the container cover tightly on the top of the cover, and then vibrate on the table 180 times, scrape the upper surface with a thin iron sheet, weigh the weight of the coal powder loaded, and then divide it by the volume of the container, which is the loose packed density.

(3) Compression ratio. The difference between the tapped bulk density and the loose bulk density is divided by the tapped bulk density, and the quotient is the compressibility. The lower the compression rate is, the harder it is for the pulverized coal to be compacted in the pulverized coal bin and tank, and the better its fluidity is.

(4) Natural slope angle. The experimental coal powder flows naturally from the height 150mm to the lower diameter 100mm pan, and then measure the angle between the slope and the horizontal plane, which is the natural slope angle.

(5) The corner of the spoon. Embed a plane scoop with a length 200mm and a width 20mm into the test coal powder pile and lift it up vertically, and measure the three-point slope angle of the coal powder on the scoop equidistantly along the length direction. Taking the arithmetic mean value is the scoop angle of the impact front plate. Vibrate the iron slider in the ladle shaft from top to bottom once, then measure the slope angles of the coal powder at three points on the ladle equidistantly along the length direction, and take the arithmetic mean value as the angle of the ladle after impact.

(6) Evenness. Arrange the tested pulverized coal in order of particle size from small to large, and the quotient of the average particle size when cumulative 60% and the average particle size when cumulative 10% is uniformity. This value reflects the uniformity of pulverized coal particles, and has a great influence on the degree of combustion of pulverized coal, because the degree of

complete combustion of pulverized coal depends on the number of coarse particles in the pulverized coal, and the combustion rate will decrease if there are more coarse particles.

2. 1. 10. 3 Characteristics

Studying the jetablility characteristics of pulverized coal includes determining the collapse angle, difference angle, dispersion and fluidity. Among them, the collapse angle, difference angle and degree of dispersion are measured through experiments. In the same experiment, each group is measured 3 times, and the average value is taken.

(1) Crash corner. After measuring the natural slope angle, use the iron slider in the pallet shaft below the plate to drop and vibrate three times naturally, and then measure the angle between the slope surface and the horizontal plane, which is the collapse angle.

(2) Poor angle. The difference between the natural slope angle and the collapse angle is called the difference angle.

(3) Dispersion. 10g experimental coal sample freely falls into the weighing pan from the height of 400mm below the weighing pan, the diameter of the weighing pan is 150mm, will splash out of the pan 10, is the degree of dispersion.

2. 1. 10. 4 Evaluation criteria and results

For the evaluation criteria of powder properties, see Table 2-19 Carr Fluidity Index Table and Table 2-20 Carr Jetness Index Table. Check the Table 2-19 from the measured value of each parameter to get each index, and sum the index to determine the fluidity evaluation of the powder. Similarly, after obtaining the measured values of each parameter, look up the Table 2-20 to get the corresponding indices, and sum the indices to determine the evaluation of the jet flow property of the powder.

Table 2-19 Carr Fluidity Index

Natural slope angle (°)		Compression rate (%)		Board Spoon Corner (°)		Evenness		Total Fluidity Index	Evaluation of fluidity
Measured values	Index	Measured values	Index	Measured values	Index	Measured values	Index		
<25	25	<5	25	<25	25	1	25	100-90	best
26-29	24	6-9	23	26-30	23	2-4	23		
30	22.5	10	22.5	31	22.5	5	22.5		
31	22	11	22	32	22	6	22	89-80	good
32-34	21	12-14	21	33-37	21	7	21		
35	20	15	20	38	20	8	20		
36	19.5	16	19.5	39	19.5	9	19.5		
37-39	18	17-19	18	40-44	18	10-11	18	79-70	
40	17.5	20	17.5	45	17.5	12	17.5		

2.1 Metallurgical properties of bituminous coal and lignite

Continued Table 2-19

Natural slope angle (°)		Compression rate (%)		Board Spoon Corner (°)		Evenness		Total Fluidity Index	Evaluation of fluidity
Measured values	Index	Measured values	Index	Measured values	Index	Measured values	Index		
41	17	21	17	46	17	13	17	69-60	ordinary
42-44	16	22-24	16	47-59	16	14-16	16		
45	15	25	15	60	15	17	15		
46	14.5	26	14.5	61	14.5	18	14.5	59-40	
47-54	12	27-30	12	62-74	12	19-21	12		
55	10	31	10	75	10	22	10		
56	9.5	32	9.5	76	9.5	23	9.5		poor
57-64	7	33-36	7	77-89	7	24-26	7	39-20	
65	5	37	5	90	5	27	5		
66	4.5	38	4.5	91	4.5	28	4.5		
67-89	2	39-45	2	92-99	2	29-35	2	19-0	worst
90	0	>45	0	>99	0	>35	0		

Table 2-20 Carr Jetness Index

Fluidity		Collapse Corner (°)		Difference angle (°)		Degree of dispersion (°)		Total Jetability Index	Evaluation of jetability
Measured values	Index	Measured values	Index	Measured values	Index	Measured values	Index		
>60	25	10	25	>30	25	>50	25	80-100	strong
59-56	24	11-19	23	29-28	24	49-44	24		
55	22.5	20	22.5	27	22.5	43	22.5		
54	22	21	22	26	22	42	22		
53-50	21	22-24	21	25	21	41-36	21		
49	20	25	20	24	20	35	20		
48	19.5	26	19.5	23	19.5	34	19.5	60-79	
47-45	18	27-29	18	22-20	18	33-29	18		
44	17.5	30	17.5	19	17.5	28	17.5		
43	17	31	17	18	17	27	17		
42-40	16	32-39	16	17-16	16	26-21	16		
39	15	40	15	15	15	20	15		
38	14.5	41	14.5	14.5	14.5	19	14.5	40-59	
37-34	12	42-49	12	12	12	18-11	12		
33	10	50	10	10	10	10	10		

Continued Table 2-20

Fluidity		Collapse Corner (°)		Difference angle (°)		Degree of dispersion (°)		Total Jetability Index	Evaluation of jetability
Measured values	Index	Measured values	Index	Measured values	Index	Measured values	Index		
32	9.5	51	9.5	9	9.5	9	9.5	25-39	↑
31-29	8	52-56	8	8	8	8	8		
28	6.25	57	6.25	7	6.25	7	6.25		
27	6	58	6	6	6	6	6	0-24	↓
26-23	3	59-64	3	5-1	3	5-1	3		
<22	0	>64	0	0	0	0	0		poor

18 types of pulverized coals fluidity and jetness properties commonly used in iron and steel enterprises are shown in Table 2-21. From the table 18 PCI coals, it can be seen from the evaluation results that most of the PCI coals have poor fluidity and medium jetness, and only a small number of coals have strong jetness, such as Baiyangshu coal, Gaoping coal and Neimenggu coal, and mainly anthracite; while the fluidity and jetness of high-volatile bituminous coal and lignite are not as good as the performance of the above-mentioned anthracite, so when the blast furnace inject bituminous coal and lignite, its transportation performance needs to be paid close attention.

Although the Carr tables apply to all pulverulent materials, pulverized coal for blast furnace injection varies not only in particle size and composition, but also in properties. Therefore, according to the situation of blast furnace pulverized coal, the difference between the fluidity and jetness index of pulverized coal should be compared, and the main factors affecting the fluidity and jetness index of pulverized coal should be analyzed.

2.1.11 Combustibility

The combustion of pulverized coal is the same as the combustion of other solid fuels, which belongs to surface combustion. Under a certain oxygen concentration and when the ambient temperature reaches above the ignition temperature of coal powder, coal powder can burn. Since the melting point of carbon is as high as 3,500℃, under normal combustion temperature, there will be no melting and sublimation, and there will be no thermal decomposition. Oxidizing gases such as O_2 diffuse to the solid surface or inside the small pores, and react with carbon on the surface of the coal powder to generate CO or CO_2 gas and diffusion from the surface to the outside world. The combustion speed of pulverized coal is very slow compared with that of gas or liquid. The combustion process includes chemical reaction, flow, heat transfer, gas diffusion, heat transfer and mass transfer inside particles, etc. It is a very complicated process.

Pulverized coal combustion includes four stages: (1) pulverized coal heats up rapidly and releases volatile matter; (2) volatile matter diffuses around and reacts with oxygen combustion of

2.1 Metallurgical properties of bituminous coal and lignite

Table 2-21 Determination results of fluidity and jetness characteristic parameters of pulverized coal

Coal	Loose Bulk density (g/m³)	Vibrate loose packing density (g/m³)	Compression ratio(%) Determination value	Compression ratio(%) Index	Natural slope angle(°) Determination value	Natural slope angle(°) Index	Spoon angle (°) Average value	Spoon angle (°) Index	Evenness Determination value	Evenness Index	Crash angle (°) Determination value	Crash angle (°) Index	Difference angle(°) Measured value	Difference angle(°) Index	Dispersion (%) Determination value	Dispersion (%) Index	Fluidity index Index	Fluidity index Index	Fluidity index Index	Describe	Jetness index	Describe
Huaren	48.00	70.00	31.43	10.00	40.00	17.50	57.50	16.00	6.82	21.00	35.00	16.00	5.00	3.00	45.00	24.00	43.50	25.00	compare difference	43.00	middle
Jiyang	40.00	58.50	31.62	9.50	55.00	10.00	56.67	16.00	4.07	23.00	45.00	12.00	10.00	10.00	50.00	25.00	35.50	24.00	difference	47.00	middle
Baiyangshu	48.00	67.00	28.36	12.00	55.00	10.00	60.00	15.00	9.66	18.00	30.00	17.50	25.00	21.00	50.00	25.00	37.00	22.50	difference	63.50	compare powerful
Lu'an	52.00	73.00	28.77	12.00	45.00	15.00	63.33	12.00	4.92	22.50	35.00	16.00	10.00	10.00	30.00	18.00	39.00	25.00	difference	44.00	middle
Guoleizhuang	44.00	61.00	27.87	12.00	45.00	15.00	65.83	12.00	7.39	21.00	25.00	20.00	20.00	18.00	35.00	20.00	39.00	25.00	difference	58.00	middle
Jingjin	42.50	60.00	29.17	12.00	45.00	15.00	73.33	12.00	3.75	23.00	40.00	15.00	5.00	3.00	30.00	18.00	39.00	25.00	difference	36.00	compare difference
Qingding	39.00	59.00	33.90	7.00	45.00	15.00	64.17	12.00	5.63	22.00	35.00	16.00	10.00	10.00	35.00	20.00	34.00	24.00	difference	46.00	middle
Bingxing	39.00	57.00	31.58	9.50	40.00	17.50	69.17	12.00	4.44	23.00	30.00	17.50	10.00	10.00	35.00	20.00	39.00	25.00	difference	47.50	middle
Sanjicun	52.00	76.00	31.58	9.50	40.00	17.50	60.00	15.00	25.18	7.00	30.00	17.50	10.00	10.00	40.00	21.00	42.00	20.00	compare difference	48.50	middle

Continued Table 2-21

Coal	Loose Bulk density (g/m³)	Vibrate loose packing density (g/m³)	Compression ratio(%) Determination value	Compression ratio(%) Index	Natural slope angle(°) Determination value	Natural slope angle(°) Index	Spoon angle(°) Average value	Spoon angle(°) Index	Evenness Determination value	Evenness Index	Crash angle(°) Determination value	Crash angle(°) Index	Difference angle(°) Measured value	Difference angle(°) Index	Dispersion(%) Determination value	Dispersion(%) Index	Fluidity index Index	Fluidity index Index	Describe	Jetness index	Describe
Xinjing	46.00	66.00	30.30	12.00	45.00	15.00	63.33	12.00	6.75	21.00	35.00	16.00	10.00	10.00	35.00	20.00	39.00	25.00	difference	46.00	middle
Fengshan	48.00	69.00	30.43	12.00	40.00	17.50	52.50	16.00	6.87	21.00	30.00	17.50	10.00	10.00	30.00	18.00	45.50	25.00	compare difference	45.50	middle
Jiaozuobei	42.00	60.00	30.00	12.00	45.00	15.00	55.00	16.00	5.91	22.00	30.00	17.50	15.00	15.00	30.00	18.00	43.00	25.00	compare difference	50.50	middle
Botou	49.50	70.50	29.79	12.00	45.00	15.00	65.83	12.00	6.45	22.00	35.00	16.00	10.00	10.00	55.00	25.00	39.00	25.00	difference	51.00	middle
Xinmeng	39.00	58.00	32.76	7.00	45.00	15.00	68.33	12.00	4.99	22.50	35.00	16.00	10.00	10.00	35.00	20.00	34.00	24.00	difference	46.00	middle
Daiwang	42.00	63.00	33.33	7.00	40.00	17.50	64.17	12.00	5.15	22.50	30.00	17.50	10.00	10.00	55.00	25.00	36.50	24.00	difference	52.50	middle
Chaoxian	47.50	72.50	34.48	7.00	40.00	17.50	65.83	12.00	4.62	22.50	35.00	16.00	5.00	3.00	25.00	16.00	36.50	24.00	difference	35.00	compare difference
Neimenggu	38.50	60.50	36.36	7.00	40.00	17.50	68.33	12.00	6.56	21.00	35.00	16.00	5.00	3.00	30.00	18.00	57.50	24.00	compare difference	61.00	stronger
Gaoping	41.50	63.50	34.65	7.00	35.00	20.00	60.00	15.00	7.17	21.00	30.00	17.50	5.00	3.00	25.00	16.00	63.00	25.00	general	61.50	stronger

CO and H_2O, volatile matter; (3) after burning of volatile matter, fixed carbon preheats and ignites; (4) residual carbon burns to completion.

The pulverized coal enters the direct blowpipe after spraying from the spray gun, and is mixed with the high-speed heat shunt, heated at a high speed, water evaporates, volatilized, decomposed, and burned. The process is extremely complicated. Generally, the pulverized coal is heated and the gasification and combustion of volatile matter are completed in the direct blowing pipe and the tuyere, and some fixed carbon starts to burn. Although the pulverized coal only stays in this area for about 5ms, this area is still an effective area for pulverized coal combustion.

The hot air leaves the tuyere at a certain angle and blows into the blast furnace to fluidize the hot coke in the furnace, forming a circulating combustion zone in the fluidized coke bed. In this area, the proportion of coke is actually not high. The combustion characteristics of pulverized coal in this area are: (1) the contact time between pulverized coal and oxygen is very short, less than 20ms; (2) the oxygen concentration in the zone is low; (3) the pulverized coal and the coke flowing in the circulation zone compete for oxygen for combustion reaction.

The combustion of pulverized coal in the raceway is more critical. If the pulverized coal cannot be burned in this area, it will be brought into the dead material column, which will affect the air permeability and liquid permeability of the blast furnace. Therefore, efforts should be made to increase the combustion rate of pulverized coal within this zone and to maintain a larger and more stable raceway combustion zone.

The combustion rate of pulverized coal is a sign of the quality of pulverized coal combustion. If the combustion rate of pulverized coal in the tuyere of the blast furnace is low and the combustion of pulverized coal is incomplete, it will not only reduce the utilization rate of pulverized coal in the blast furnace, but also affect the gas permeability of the charge, thereby affecting the production of the blast furnace. Therefore, for blast furnace coal injection, the pulverized coal combustion rate is an important indicator to measure the performance of pulverized coal and judge the combustion status of the tuyere of the blast furnace. Enhancing the combustion of pulverized coal at the tuyere is the most important factor for the large injection of the blast furnace.

The combustibility of pulverized coal in blast furnace is usually determined by microcomputer differential thermal balance. When doing thermogravimetric experiments, it is first necessary to accurately weigh the weight of the sample, then add a certain amount of pulverized coal into the ground snail of the sample, place it on a differential thermal balance, and pass a certain amount of air flow (60mL/min), according to a certain heating rate (20℃/min) heat pulverized coal. As the temperature increases, the pulverized coal is first heated rapidly, followed by degassing and rapid thermal decomposition (that is, the thermal decomposition of coal and the secondary decomposition of volatiles), and then ignited, the volatiles are burned, and finally the residual carbon (or semi-coke) and oxygen for multi-phase combustion reaction until the pulverized coal is completely burned. Use thermal analysis to determine the combustion rate of pulverized coal to 500℃, 600℃, 700℃.

(1) Experimental method. A certain amount of pulverized coal is loaded into the differential

heat meter, and the air flow is introduced to heat the pulverized coal at a heating rate. The pulverized coal is heated to complete combustion. From the weight loss curve (TG) find the weight percentage of pulverized coal burning to 500℃, 600℃, 700℃. The combustion rate is the total flammable value (that is, all weight loss percentages after complete combustion of pulverized coal).

(2) Experimental parameters. Sample weight about 18.0mg, air flow 60mL/min, heating rate 20℃/min.

See Table 2-22 for the measurement results of combustibility index of PCI coal commonly used in iron and steel enterprises, and comparative analysis chart is shown in Fig. 2-12.

Table 2-22 **Combustibility index of PCI coal commonly used in iron and steel enterprises**

Sample serial number	Type	Pulverized coal	500℃ Combustion rate(%)	600℃ Combustion rate(%)	700℃ Combustion rate(%)
1	anthracite	Qinding 1	8.13	47.5	83.77
2	anthracite	Jiaozuobei	13.88	37.32	73.75
3	anthracite	Botou	6.77	41.37	81.09
4	anthracite	Haolin	3.52	36.2	74.3
5	anthracite	Poplar Villa	1.74	31.48	72.75
6	anthracite	Russian coal	2.86	22.18	69.7
7	anthracite	Dongpei	31.8	72.56	99.7
8	anthracite	Guanghui	31.83	75.75	99.28
9	anthracite	Ningmei	8.55	55.06	98.84
10	anthracite	Zhongxin	45.75	86.69	99.72
11	lean coal	Sanji Village	19.42	56.2	91.27
12	lean coal	Lu'an	15.26	49.12	83.48
13	bituminous coal	magical powers	59.16	88.69	99.6
14	bituminous coal	Xuanhua	59.71	88.43	100
15	bituminous coal	Gonghong	57.07	83.89	100
16	gas coal	Jiyang	42.29	77.52	99.96
17	lignite	Zincai	80.53	99.19	100
18	lignite	New energy	61.89	85.09	100
19	lignite	Binxin	78.54	98.84	99.97

Based on the above analysis, it can be found that the combustibility of different coal types is significantly different, and the combustibility gradually decreases with the increase of coal metamorphism. That is, the combustion rate of anthracite, bituminous coal, and lignite under the same conditions increase in turn. From this analysis, it can be concluded that when the blast furnace is injected with lignite and bituminous coal, it can significantly improve the combustion effect of pulverized coal in the tuyere and raceway swirl area and increase the combustion rate of pulverized coal.

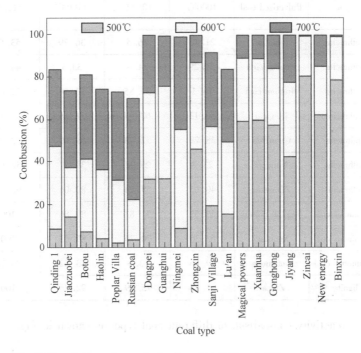

Fig. 2-12 Comparative analysis of combustibility of different coal types

2.1.12 Reactivity

After the amount of coal injection reaches a certain level, a large amount of pulverized coal residue will appear in the blast furnace, which will deteriorate the gas permeability of the charge and affect the operation of the blast furnace. A small part of the unburned pulverized coal residue is consumed by direct reduction in the blast furnace, and most of it is consumed by $C+CO_2 \rightarrow 2CO$ reaction consumption. The chemical reactivity of coal to CO_2 refers to the ability of carbon in coal to undergo reduction reaction with CO_2 at a certain temperature, or coal have the ability of CO_2 to be reduced to CO, the amount of CO_2 reduced to CO accounts for the total amount of CO_2 participating in the reaction expressed as a percentage of the amount $\alpha(\%)$. Also commonly referred to as the reactivity of coal to CO_2.

During the experiment, about 18.0mg coal powder was put into the differential thermal balance, and oxygen (20mL/min) was passed through for protection, and 20℃/min heating up to 900℃, drying to remove moisture and volatiles. Then feed CO_2(20mL/min) to react with the sample, and start to record the weight loss of the sample, and raise the temperature to (20℃/min) 1,200℃. So far, the calculated weight loss rate at a certain temperature is the reactivity.

See Table 2-23 for the gasification reactivity index of PCI coal commonly used in iron and steel enterprises.

Table 2-23 Combustibility index of PCI coal commonly used in iron and steel enterprises

Sample serial number	Type	Pulverized coal	1000℃	1050℃	1100℃	1150℃	1200℃
1	anthracite	Haolin	21.16	26.5	36.79	53.54	76.56
2	anthracite	Poplar Villa	21.29	24.4	33.3	44.94	68.23
3	anthracite	Russian coal	16.1	21.85	33.54	52.09	76.21
4	anthracite	Dongpei	50.13	60.94	71.56	81.13	91.19
5	anthracite	Guanghui	38.29	52.78	69.09	83.78	95.95
6	anthracite	Ningmei	23.8	29.38	39.71	52.64	67.74
7	anthracite	Zhongxin	40.95	58.24	80.06	96.58	99.93
8	bituminous coal	Magical powers	56.55	71.81	89.86	100	100
9	bituminous coal	Xuanhua	55.36	70.76	89.03	100	100
10	bituminous coal	Gonghong	49.85	63.9	83.72	99.38	100
11	lignite	Zincai	61.75	75.99	92.76	100	100

The gasification reactivity comparison of different coal types is shown in Fig. 2-13.

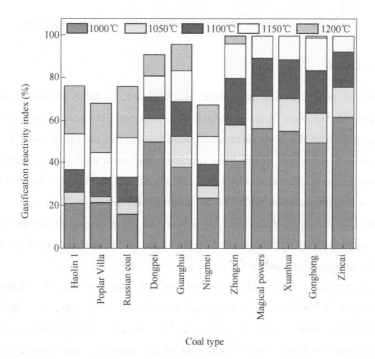

Fig. 2-13 Comparative analysis of gasification reactivity of different coal types

Based on the above analysis, it can be found that with the increase of coal metamorphism, the gasification reactivity of blast furnace injection coal gradually improves, that is, under the same conditions, the gasification reaction index of anthracite, bituminous coal, and lignite gradually increases. Therefore, compared with anthracite, the unburned dust generated when lignite and bituminous coal are injected into blast furnaces is easier to consume, which has a better advantage in ensuring the gas permeability of the furnace charge.

2.2 Interaction law of mixed injection of bituminous coal, lignite and anthracite

2.2.1 The influence of bituminous coal and lignite on the basic performance of blended coal

Based on the proximate analysis of coal types, volatile content, ash content, and sulfur content are selected as the basis for coal blending, and volatile content is the most important reference factor. Researches at home and abroad have shown that the volatile content of mixed coal at 13%-20% is the best combination, which can achieve the highest pig iron production, the highest coal-coke replacement ratio, and the best pig iron quality. Coal blending is carried out around 18%, 19%, 20%. In order to expand the scope of PCI coal resources, the volatile content of blended coal was increased in the coal blending experiment scheme, and the highest volatile content was 30.53%. In addition, control the ash content of blended coal to be less than 12%, sulfur content to be less than 0.6% to determine the coal blending scheme. Since the composition of mixed coal (volatile matter, ash, fixed carbon, sulfur, moisture, etc.) is equal to the weighted average of each mixed coal type, it can be calculated according to the proximate analysis and ultimate analysis results of existing coal types to determine (Table 2-24) for eligible coal blending schemes. In the experimental scheme, Jinkai coal is a typical lignite coal, Xuanhua coal is a typical high-volatile bituminous coal, and Baiyangshu coal is a typical anthracite coal.

Table 2-24 Basic properties of coal blends

Serial number	Matching plan	Cost (yuan)	Ash (%)	S content (%)	C content (%)	Volatile matter (%)
1	60%Xuanhua+40%Poplar Villa	937	9.9	0.514	73.938	23.012
2	70%Xuanhua+30%Poplar Villa	909	9.85	0.472	72.741	25.634
3	80%Xuanhua+20%Poplar Villa	881	9.8	0.432	71.544	28.256
4	10%Jinkai+50%Xuanhua+40%Baiyangshu	962	9.817	0.498	73.099	23.77
5	10%Jinkai+60%Xuanhua+30%Baiyangshu	934	9.767	0.457	71.902	26.39
6	10%Jinkai+70%Xuanhua+20%Baiyangshu	906	9.717	0.417	70.705	29.014
7	20%Jinkai+20%Xuanhua+60%Baiyangshu	1024	9.83	0.56	74.65	19.284
8	20%Jinkai+30%Xuanhua+50%Baiyangshu	1014	9.78	0.52	73.45	21.906

Continued Table 2-24

Serial number	Matching plan	Cost (yuan)	Ash (%)	S content (%)	C content (%)	Volatile matter (%)
9	20%Jinkai+40%Xuanhua+40%Baiyangshu	986	9.73	0.48	72.26	24.528
10	20%Jinkai+50%Xuanhua+30%Baiyangshu	958	9.68	0.44	71.063	27.15
11	20%Jinkai+60%Xuanhua+20%Baiyangshu	930	9.634	0.4	69.866	29.772
12	30%Jinkai+30%Xuanhua+40%Baiyangshu	1011.4	9.65	0.466	71.42	25.286
13	30%Jinkai+40%Xuanhua+30%Baiyangshu	983.5	9.6	0.425	70.224	27.908
14	30%Jinkai+50%Xuanhua+20%Baiyangshu	955	9.551	0.384	69.027	30.53

With the increase of the proportion of bituminous coal, the content of volatile matter in the blended coal increases and the content of carbon decreases.

Xuanhua coal has the lowest price, so as the proportion of Xuanhua coal increases, the price of mixed coal gradually decreases. Under the condition of scheme 3 (80%Xuanhua+20%Baiyangshu), The price of blended coal is the lowest.

The S content in Baiyangshu is higher, therefore, with the decrease of Baiyangshu content, the S content in the blended coal decreases continuously.

Since the ash content in Jinkai coal is the lowest, as the content of Jinkai coal increases, the ash content gradually decreases.

2.2.2 Effects of bituminous coal and lignite on ignition point and explosiveness of mixed coal

Using the aforementioned method for determining the ignition point, explosiveness, and ignition point of a single coal powder, use the ignition point measuring device and the long-tube type pulverized coal explosiveness testing equipment to measure the ignition point and explosiveness of 14 kinds of coal blends, and use an automatic calorimeter to measureits calorific value, the measurement results are shown in the Table 2-25.

Table 2-25 Ignition point, explosiveness and low calorific value of mixed coal

Serial number	Experimental scheme	Ignition point(℃)	Return flame(mm)	Low heating value(J/g)
1	60%Xuanhua+40%Poplar Villa	332.25	15	28,442.16
2	70%Xuanhua+30%Poplar Villa	324.00	16	27,985.22
3	80%Xuanhua+20%Poplar Villa	316.00	40	27,528.28
4	10%Jinkai+50%Xuanhua+40%Baiyangshu	324.00	0	28,089.71
5	10%Jinkai+60%Xuanhua+30%Baiyangshu	320.00	0	27,632.77
6	10%Jinkai+70%Xuanhua+20%Baiyangshu	321.67	30	27,175.89
7	20%Jinkai+20%Xuanhua+60%Baiyangshu	370.00	0	28,651

2.2 Interaction law of mixed injection of bituminous coal, lignite and anthracite

Continued Table 2-25

Serial number	Experimental scheme	Ignition point(℃)	Return flame(mm)	Low heating value(J/g)
8	20%Jinkai+30%Xuanhua+50%Baiyangshu	331.00	0	28,194
9	20%Jinkai+40%Xuanhua+40%Baiyangshu	325.00	0	27,737
10	20%Jinkai+50%Xuanhua+30%Baiyangshu	317.33	25	27,280
11	20%Jinkai+60%Xuanhua+20%Baiyangshu	310.33	0	26,823.39
12	30%Jinkai+30%Xuanhua+40%Baiyangshu	330.33	0	27,384.8
13	30%Jinkai+40%Xuanhua+30%Baiyangshu	325.33	5	26,927.88
14	30%Jinkai+50%Xuanhua+20%Baiyangshu	310.33	0	26,470.94

Experimental detection found that the general trend of the ignition point of mixed coal is that it gradually decreases with the increase of bituminous coal ratio. After adding anthracite to bituminous coal, the explosiveness of the mixed pulverized coal decreases rapidly. Among the 14 schemes, the highest ratio of bituminous coal to anthracite reaches 4 : 1, the volatile content of 30%, is still not explosive, which shows that the addition of anthracite will quickly reduce the explosiveness of bituminous coal.

The overall change trend of the calorific value of the blended coal also gradually decreases with the increase of the proportion of bituminous coal, especially when the proportion of Jinkai lignite increases, the calorific value of the blended coal decreases greatly.

2.2.3 Effect of bituminous coal and lignite on combustibility of blended coal

According to the coal blending plan, use differential thermal analysis to measure the combustion rate of various mixed coal powders to 500℃, 600℃ and 700℃, The results are shown in Table 2-26.

Table 2-26 Combustion rate of coal blends at different temperatures (%)

Serial number	Experimental scheme	500℃	600℃	700℃
1	60%Xuanhua+40%Poplar Villa	47.89	79.35	99.75
2	70%Xuanhua+30%Poplar Villa	49.8	77.90	99.71
3	80%Xuanhua+20%Poplar Villa	51.4	81.3	99.82
4	10%Jinkai+50%Xuanhua+40%Baiyangshu	45.21	75.34	99.73
5	10%Jinkai+60%Xuanhua+30%Baiyangshu	51.76	81.56	99.21
6	10%Jinkai+70%Xuanhua+20%Baiyangshu	53.00	82.60	100
7	20%Jinkai+20%Xuanhua+60%Baiyangshu	33.21	65.16	98.96
8	20%Jinkai+30%Xuanhua+50%Baiyangshu	39.75	69.22	99.52
9	20%Jinkai+40%Xuanhua+40%Baiyangshu	44.00	73.91	99.65
10	20%Jinkai+50%Xuanhua+30%Baiyangshu	43.70	69.73	95.99

Continued Table 2-26

Serial number	Experimental scheme	500℃	600℃	700℃
11	20%Jinkai+60%Xuanhua+20%Baiyangshu	59.18	86.75	99.90
12	30%Jinkai+30%Xuanhua+40%Baiyangshu	46.44	73.69	99.13
13	30%Jinkai+40%Xuanhua+30%Baiyangshu	49.05	77.74	100
14	30%Jinkai+50%Xuanhua+20%Baiyangshu	61.05	88.49	99.96

According to the measurement results, the combustion rate changes of 14 kinds of mixed coals burned to 500℃, 600℃, 700℃, curve as shown in Fig. 2-14. The general trend of the combustibility of the blended coal is that it gradually increases with the increase of the bituminous coal ratio; especially with the increase of the ratio of the best combustible Jinkai lignite, the combustibility of the blended coal increases even more. Therefore, the addition of Jinkai lignite helps to improve the combustibility of blended coal.

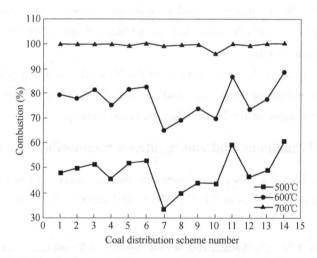

Fig. 2-14 Combustion performance relationship curve of mixed coal

2.2.4 Effects of bituminous coal and lignite on gasification reactivity of blended coal

The reaction performance of 14 coal blends was characterized by measuring the reaction conversion at 1,000℃, 1,050℃ and 1,100℃ according to the coal blending scheme using differential thermal analysis, the measurement results are shown in the Table 2-27.

Table 2-27 Reaction conversion rate of mixed coal at different temperatures (%)

Serial number	Experimental scheme	1000℃	1050℃	1100℃
1	60%Xuanhua+40%Poplar Villa	45.85	57.87	73.42
2	70%Xuanhua+30%Poplar Villa	46.76	59.33	73.85
3	80%Xuanhua+20%Poplar Villa	49.58	63.57	78.59
4	10%Jinkai+50%Xuanhua+40%Baiyangshu	44.52	55.55	67.07

2.2 Interaction law of mixed injection of bituminous coal, lignite and anthracite

Continued Table 2-27

Serial number	Experimental scheme	1000℃	1050℃	1100℃
5	10%Jinkai+60%Xuanhua+30%Baiyangshu	47.50	60.61	74.05
6	10%Jinkai+70%Xuanhua+20%Baiyangshu	51.63	64.43	78.96
7	20%Jinkai+20%Xuanhua+60%Baiyangshu	42.44	50.91	59.69
8	20%Jinkai+30%Xuanhua+50%Baiyangshu	41.83	51.19	61.57
9	20%Jinkai+40%Xuanhua+40%Baiyangshu	44.71	55.98	67.43
10	20%Jinkai+50%Xuanhua+30%Baiyangshu	49.74	61.82	72.84
11	20%Jinkai+60%Xuanhua+20%Baiyangshu	51.30	63.77	78.37
12	30%Jinkai+30%Xuanhua+40%Baiyangshu	49.25	60.65	71.32
13	30%Jinkai+40%Xuanhua+30%Baiyangshu	52.23	64.87	76.58
14	30%Jinkai+50%Xuanhua+20%Baiyangshu	53.98	67.25	81.36

According to the measurement results, 14 mixed coals react with CO_2 to 1,000℃, 1,050℃, 1,100℃, the relationship curve between temperature points and conversion rate is shown in Fig. 2-15. It can be found that the overall change trend of the reactivity of the blended coal is gradually enhanced with the increase of the bituminous coal ratio; especially with the increase of the most reactive Jinkai lignite ratio, the reactivity of the blended coal increases faster. Therefore, the addition of Jinkai lignite helps to improve the reactivity of the blended coal.

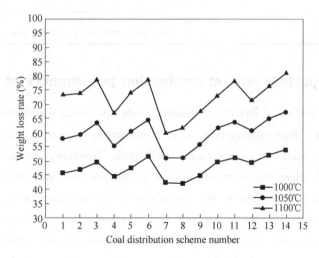

Fig. 2-15 Coal blended reaction curve

2.2.5 Research on optimization of coal blending scheme

It can be seen from Table 2-24 that when Xuanhua bituminous coal is mixed with Baiyangshu anthracite, the scheme 3 is the most economical scheme, and has the best combustibility and reactivity without explosiveness. However, the ignition point is lower and the volatile content is higher. Compared with the combination of Jinkai lignite, Xuanhua bituminous coal and Baiyangshu anthracite, the cost performance of Xuanhua bituminous coal and Baiyangshu anthracite is lower. Among the Jinkai lignite with 10%, scheme 5 is the best combustibility scheme, and the volatile matter content is 26%, blast furnace is easy to operate accepted, better reactivity and better security.

Due to the low price of Jinkai lignite, proper blending of Jinkai lignite can not only increase the volatile content of the blended coal, but also improve the combustibility of the blended coal. From the results of coal blending experiments, it can be seen that when the volatile content of blended coal is 28%, the scheme with the best combustibility and reactivity should be selected 13; when the volatile content of the blended coal is 30%, the scheme with better combustibility and reactivity should be selected 11, Although scheme 14 has the best flammability and reactivity, but scheme 14 has the lowest ignition point and calorific value. From the analysis of operation safety, choose scheme 11 would be appropriate.

Based on the above experimental analysis results, four schemes in Table 2-28 are drawn up for on-site industrial experiments.

Table 2-28 Recommended coal blending scheme

Serial number	Experimental scheme
3	80%Xuanhua+20%Poplar Villa
5	10%Jinkai+60%Xuanhua+30%Baiyangshu
11	20%Jinkai+60%Xuanhua+20%Baiyangshu
13	30%Jinkai+40%Xuanhua+30%Baiyangshu

2.3 Effect of particle size on combustion performance of low-rank coal

In recent years, domestic and foreign ironmaking plants have found that relaxing coal particle size can save energy and reduce consumption, reduce coal production cost, and increase pulverizing capacity through experiments on increasing coal ratio. However, excessive particle size of pulverized coal will affect the combustibility of pulverized coal and increase the amount of unburned pulverized coal, which is not conducive to the utilization rate of pulverized coal. Therefore, the efficient utilization of low-rank coal injection in blast furnaces requires systematic research on the combustibility of low-rank coal at different particle sizes, and then determines the appropriate injection particle size. In this section, a differential thermal balance is used to study the combustibility of different coal types at different particle sizes.

2.3.1 Effect of particle size on combustibility of anthracite

Baiyangshu pulverized coal screen was divided into four grades: 50-100 mesh, 100-140 mesh, 140-200 mesh, -200 mesh (50 mesh = 270μm, 100 mesh = 150μm, 140 mesh = 109μm, 200 mesh = 75μm, the same below), and the differential thermal balance was used to carry out combustion test. The combustion rate results at different temperatures were obtained by analyzing the data, as shown in Table 2-29.

Table 2-29 Combustibility of Baiyangshu pulverized coal with different particle sizes (%)

Particle size (mesh)	500℃	550℃	600℃	650℃	700℃	750℃	800℃	850℃	900℃
50-100	0.11	4.38	18.02	37.22	57.28	75.00	91.52	100	100
100-140	1.80	8.61	24.70	45.66	66.26	83.88	98.89	100	100
140-200	1.82	9.57	29.04	52.74	73.09	85.86	99.38	100	100
-200	1.84	12.80	31.48	53.08	72.76	88.71	99.89	100	100

In order to analyze the change law of pulverized coal combustion rate with particle size more clearly, a figure is drawn based on the data in Table 2-23, and the results are shown in Fig. 2-16. As can be seen from Fig. 2-16, there is a good correlation between Baiyangshu pulverized coal flammability and particle size, that is, the smaller the particle size, the better the flammability. The temperature region where the flammability varies most obviously with particle size is 600-800℃. When the maximum combustion temperature is more than 800℃ and the particle size is less than 0.15mm (100 mesh), the pulverized coal can be fully burned. When the highest combustion temperature is 700℃, the particle size should be less than 0.106mm (140 mesh), in order to achieve a better combustion effect.

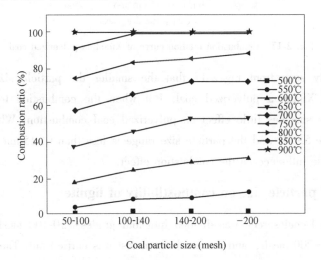

Fig. 2-16 Coal powder combustion curve in Baiyangshu

2.3.2 Effect of particle size on the combustibility of bituminous coal

Xuanhua pulverized coal screen was divided into four grades: 50-100 mesh, 100-140 mesh, 140-200 mesh and -200 mesh, and combustion performance test was conducted. The data are shown in Table 2-30.

Table 2-30 Combustibility of Xuanhua pulverized coal with different particle sizes (%)

Particle size(mesh)	500℃	550℃	600℃	650℃	700℃	750℃	800℃	850℃	900℃
50-100	52.00	70.63	86.94	98.16	98.70	99.52	100	100	100
100-140	62.79	79.75	94.05	99.19	99.59	100	100	100	100
140-200	62.91	80.73	95.92	99.27	99.67	100	100	100	100
-200	69.71	81.14	96.43	99.42	100	100	100	100	100

Fig. 2-17 shows the result obtained after analyzing the data in Table 2-30.

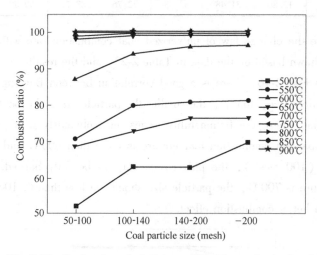

Fig. 2-17 Combustion relation curve of Xuanhua pulverized coal

It can be clearly seen from Fig. 2-17 that the smaller the particle size, the higher the combustion rate of Xuanhua pulverized coal. But when the combustion temperature exceeds 650℃, the particle size has little effect on pulverized coal combustion. When the combustion temperature is above 550℃ and the particle size range is less than 0.15mm (100 mesh), the particle size has little influence on the combustion effect.

2.3.3 Effect of particle size on combustibility of lignite

The Jinkai pulverized coal screen was divided into four grades: 50-100 mesh, 100-140 mesh, 140-200 mesh and -200 mesh, and the combustion test was carried out. The data are shown in Table 2-31.

Table 2-31 Flammability of Jinkai pulverized coal with different particle sizes

Particle size (mesh)	500℃	550℃	600℃	650℃	700℃	750℃	800℃	850℃	900℃
50-100	70.94	85.07	97.28	99.86	100	100	100	100	100
100-140	72.33	85.37	96.51	99.37	99.97	100	100	100	100

Continued Table 2-31

Particle size (mesh)	500℃	550℃	600℃	650℃	700℃	750℃	800℃	850℃	900℃
140-200	73.13	85.86	96.64	98.99	99.67	100	100	100	100
-200	80.52	92.05	99.19	99.38	100	100	100	100	100

Fig. 2-18 shows the result obtained after analyzing the data in Table 2-31.

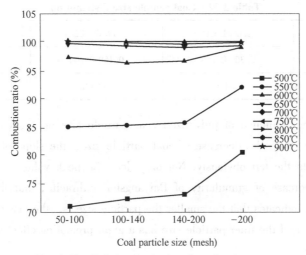

Fig. 2-18 Jinkai pulverized coal combustion curve

As can be seen from Fig. 2-18, when the combustion temperature is below 600℃, the combustion rate increases gradually with the decrease of particle size. When the combustion temperature is higher than 600℃, the particle size of Jinkai pulverized coal has little influence on the pulverized coal combustion effect, and the combustion rate is higher than 95% under any particle size condition.

It is found that the combustion effect of bituminous coal is obviously higher than that of anthracite by comprehensive analysis of the combustion characteristic curves of the three kinds of pulverized coal. When the combustion temperature is 600℃, the combustion rate of anthracite with -200 mesh (less than 0.074mm) particle size is only 31%, while that of bituminous coal with the same particle size is 96%, and Jinkai coal is almost completely burned out. Under the same combustion conditions, bituminous coal particle size can be slightly coarser, anthracite particle size can be slightly finer. From the data analysis, it can be seen that the granularity of bituminous coal should be controlled within 100 mesh, while that of anthracite coal should be less than 140 mesh, and the particle size is mainly -200 mesh.

2.3.4 Combustion characteristics and kinetic analysis of pulverized coal with different particle size

2.3.4.1 Combustion thermogravimetric curve analysis

Fig. 2-19-Fig. 2-21 show the thermal weight loss (TG) and weight loss differential (DTG) curves of three kinds of pulverized coal with different degrees of metamorphism in different particle size ranges. TG curve is the curve of coal sample mass changing with temperature, while DTG curve is the instantaneous weight loss rate calculated according to TG curve, indicating the severity of weight loss at a certain moment. A, B and C represent Baiyangshu anthracite, Xuanhua bituminous coal and Jinkai lignite, respectively. The particle size distribution of coal samples is shown in Table 2-32.

Table 2-32 Coal sample size distribution

Coal sample	A1/B1/C1	A2/B2/C2	A3/B3/C3
Size distribution (mm)	0.300-0.150	0.150-0.106	0.106-0.075
Number range	50-100	100-140	140-200

Fig. 2-19 shows the influence of pulverized coal particle size on the combustion process of Baiyangshu anthracite. With the decrease of coal particle size, the slope of TG curve increases gradually and moves to the left obviously. Not only does the peak value of DTG curve increase gradually with the decrease of granularity of Baiyangshu anthracite, but the whole curve also shifts to the left. This indicates that the smaller the particle size is, the faster the combustion rate of pulverized coal is, and the finer particle size has a great promoting effect on the combustion of Baiyangshu anthracite.

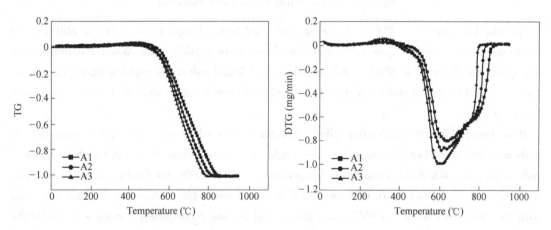

Fig. 2-19 TG and DTG curves of Baiyangshu anthracite combustion

Fig. 2-20 shows the influence of pulverized coal particle size on the combustion process of Xuanhua bituminous coal. With the decrease of particle size, the slope of TG curve gradually increases and moves to the left. The peak value of DTG curve increases gradually, and the curve moves to the left gradually. However, within the experimental range, the smaller the particle size of Xuanhua bituminous coal, the smaller the change amplitude of TG curve, and it can be observed that there is not much difference between 100-140 mesh and 140-200 mesh of Xuanhua bituminous coal DTG curve.

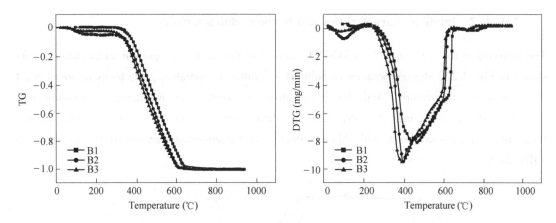

Fig. 2-20　TG and DTG curves of Xuanhua bituminous coal combustion

Fig. 2-21 shows the influence of particle size on the combustion process of Jinkai lignite. Within the experimental range, granularity refinement has little influence on TG and DTG curves of Jinkai lignite, indicating that granularity has little influence on the combustion of Jinkai lignite within the experimental range. The TG and DTG curves of Jinkai lignite are obviously different from those of anthracite and bituminous coal: TG has two obvious steps and DTG curve has two peaks. In the initial stage (about 200℃ before), the moisture and volatilization of pulverized coal are analyzed, which is the first step in the TG curve and the first peak in the DTG curve. After about 300℃, the volatile matter and fixed carbon in pulverized coal begin to burn, which is manifested as the second step in the TG curve and the second peak in the DTG curve.

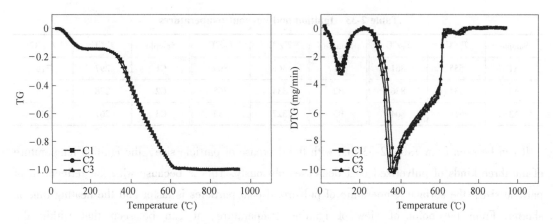

Fig. 2-21　TG and DTG curves of Jinkai lignite combustion

As can be seen from Fig. 2-19 to Fig. 2-21, pulverized coal particle size refinement has different effects on promoting combustion of pulverized coal with different metamorphism degrees. Pulverized coal with high metamorphism and smaller particle size has a good effect on promoting pulverized coal combustion, such as Baiyangshu anthracite. However, for coal powder with low metamorphism, such as Jinkai lignite, within the experimental range (50-200 mesh), the change of particle size has little influence on its combustion.

2.3.4.2 Ignition characteristics and burnout characteristics

The following is the commonly used TG-DTG method to determine the ignition temperature[2]. As shown in Fig. 2-22, the temperature at point M is ignition temperature, the temperature at point N is burnout temperature, and the temperature at point L is maximum combustion rate temperature. By analyzing the experimental data, the ignition temperature and burnout temperature of pulverized coal with different degrees of metamorphism are obtained, as shown in Table 2-33.

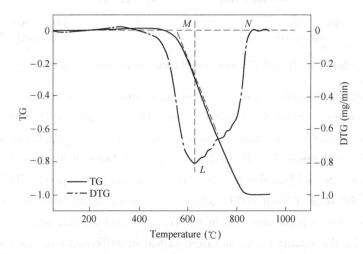

Fig. 2-22 Determination of ignition temperature by TG-DTG method

Table 2-33 Ignition and burnout temperatures

Sample	$T_i/℃$	$T_F/℃$	Sample	$T_i/℃$	$T_F/℃$	Sample	$T_i/℃$	$T_F/℃$
A1	555	861	B1	366	809	C1	297	718
A2	534	836	B2	334	765	C2	278	735
A3	518	804	B3	327	753	C3	267	743

It can be seen from Table 2-33 that with the decrease of particle size, the ignition temperature of the three kinds of pulverized coal decreases obviously. This is because with the refinement of particle size, the surface temperature of pulverized coal particles is faster and the heating time is shorter. From the point of view of ignition temperature, it can be seen that within the experimental range, smaller particle size has roughly the same effect on ignition points of pulverized coal with different degrees of metamorphism, that is, it can significantly reduce the ignition temperature of pulverized coal, which is meaningful to promote combustion.

It can also be seen from Table 2-33 that with the reduction of coal particle size, the burnout temperature of Baiyangshu anthracite and Xuanhua bituminous coal decreases gradually. However, with the decrease of particle size, the decreasing range of burntout temperature of Xuanhua bituminous coal becomes smaller. Jinkai lignite shows the opposite rule. With the refinement of particle size, the burnout temperature of Jinkai lignite increases gradually. This is because Jinkai lignite is a low-metamorphic coal, and the large pores in its interior play an important role in the combustion reaction of organic matter. With the granularity refinement of Jinkai lignite, the large pores in pulverized coal particles are destroyed, resulting in the reaction rate of pulverized coal at the later stage of combustion is not as fast as that of large particles, and the burningout time of fine pulverized coal is longer than that of coarse particles. Therefore, with the decrease of particle size, the burnout temperature of Jinkai lignite increases gradually. From the point of view of burnout temperature, it can be concluded that the smaller particle size promotes the combustion of pulverized coal with high metamorphism in the experimental range. But for the combustion of pulverized coal with low metamorphism, its particle size should not be too fine, which is not conducive to promoting combustion.

2.3.4.3 Composite combustion characteristics

The comprehensive combustion characteristic index S_N[3] was used to evaluate the combustion characteristics of coal samples. The comprehensive combustion characteristic index S_N represents the comprehensive combustion performance of coal. The higher the S_N value, the better the combustion characteristics of coal.

$$S_N = \frac{\left(\frac{dG}{d\tau}\right)_{max} \left(\frac{dG}{d\tau}\right)_{mean}}{T_i^2 T_F} \tag{2-17}$$

where $(dG/d\tau)_{max}$ —— Maximum combustion rate, mg/min;
  $(dG/d\tau)_{mean}$ —— Mean combustion rate, mg/min;
  T_F —— Burnout temperature, ℃;
  T_i —— Ignition temperature, ℃.

Table 2-34 shows the calculation results of test samples.

Table 2-34 Comprehensive combustion characteristic index of pulverized coal combustion

Sample	$(dG/d\tau)_{max}$ (mg/min)	$(dG/d\tau)_{mean}$ (mg/min)	S_N (mg^2/(min$^2 \cdot$ ℃^{3}))
A1	0.8050	0.5770	1.75×10^{-9}
A2	0.8779	0.6409	2.36×10^{-9}
A3	0.9963	0.5724	2.64×10^{-9}
B1	0.8208	0.3847	2.91×10^{-9}
B2	0.9709	0.4302	4.89×10^{-9}
B3	0.9820	0.4285	5.23×10^{-9}
C1	1.0130	0.4238	7.96×10^{-9}
C2	1.0180	0.4096	7.34×10^{-9}
C3	1.0194	0.4050	7.79×10^{-9}

As can be seen from Table 2-34, the comprehensive combustion characteristic index S_N of the three kinds of pulverized coal has the same variation law with particle size. With the decrease of pulverized coal particle size, the comprehensive combustion characteristic index S_N increases gradually. Compared with Baiyangshu anthracite and Xuanhua bituminous coal, the comprehensive combustion characteristic index of Jinkai lignite varies much less with the grain size.

Combined with the influence of pulverized coal particle size on ignition characteristics and burnout temperature, the variation law of comprehensive combustion characteristic index of pulverized coal with different metamorphism can be explained. Because of the low volatile content and high fixed carbon content of pulverized coal with high metamorphism, there are few micropores and thermal decomposition substances in raw coal particles, so the combustion is mainly stratified. When the particle size is refined, the pulverized coal particles become smaller, the heat transfer rate becomes faster, the ignition time of pulverized coal is advanced, the ignition temperature is reduced, the maximum combustion rate is increased accordingly, and the final burnout temperature is also decreased. Therefore, the particle size change has a great influence on the combustion characteristics of the high-metamorphic pulverized coal (Baiyangshu anthracite) within the experimental range (50-200 mesh).

Low metamorphic pulverized coal is characterized by high volatile content, high internal water content and relatively little fixed carbon content. Raw coal is porous and loose in structure, and combustion is dominated by internal pores. In the combustion process, large pores play an important role in the combustion reaction of internal organic matter, and it is the main channel for the diffusion of reactive gas into and out of the particles. When the particle size is refined, the pulverized coal particles become smaller, the heat transfer rate becomes faster, the volatiles and moisture are precipitated rapidly, the volatiles of pulverized coal are ignited in advance, and the ignition temperature is reduced. However, due to the destruction of the macroporous structure in the raw coal, the combustion rate of fixed carbon in pulverized coal decreases and the burnout temperature increases. Therefore, the change of particle size has little effect on the combustion characteristics of pulverized coal with low metamorphism under experimental conditions.

2.3.4.4 Combustion kinetics

Assuming that in the infinitesimal time interval, non-isothermal process can be regarded as isothermal process, according to the Arrehenius equation and mass action law, non-isothermal thermogravimetric experiment reaction rate equation can be expressed as follows[4]:

$$\frac{d\alpha}{dT} = \frac{A}{\beta} A \exp(-E_a/RT)(1-\alpha)^n \tag{2-18}$$

Sample conversion rate α can be obtained by TG curve.

$$\alpha = \frac{m_0 - m_t}{m_0 - m_\infty} \tag{2-19}$$

Where, β is the heating rate; A is the pre-exponential factor; E_a is the activation energy of the reaction; $R = 8.314 \text{J}/(\text{mol} \cdot \text{K})$, R is the gas constant; n is the order of reaction; m_0, m_t, and m_∞ represent the weight of the sample before the reaction, at time t, and at the end of the reaction, respectively.

In this section, the Coats-Redferns integral method was used to calculate TG and DTG data of samples. Reaction order $n=1$ was selected for pulverized coal combustion, and the approximate solution was obtained:

$$\ln\left[\frac{-\ln(1-\alpha)}{T^2}\right] = \ln\left[\frac{AR}{\varphi E}\left(1 - \frac{2RT}{E}\right)\right] - \frac{E}{RT} \quad (2\text{-}20)$$

Let the left side of Eq. (2-19) be Y, and the right $1/T$ be X. Then, the activation energy E can be calculated according to the slope by drawing A graph. Then, the value of E can be substituted into the intercept to get the value of frequency factor A.

It can be seen from the TG and DTG curves of pulverized coal that the curves change differently at different stages of pulverized coal combustion. In this section, the weighted average apparent activation energy proposed by J. W. Camming is used to evaluate the reactivity of pulverized coal[5]:

$$E_m = E_1 f_1 + E_2 f_2 + \cdots + E_n f_n \quad (2\text{-}21)$$

Where, E_1-E_n is the apparent activation energy of each reaction stage; $f_1 \sim f_n$ is the percentage of weight loss in total weight loss at each reaction stage.

Table 2-35 shows the results of solving the average apparent activation energy of each coal sample according to this method.

Table 2-35 Average apparent activation energy of each coal sample

Coal sample	Temperature range (℃)	Weight loss share (%)	Apparent activation energy of each stage E_i(kJ/mol)	Coefficient of association R	Average apparent activation energy E_m (kJ/mol)
A1	531-609	16.18	144.828	0.9835	54.443
	609-805	74.99	34.770	0.9949	
	805-858	6.73	73.343	0.9895	
A2	485-605	25.92	59.826	0.9967	49.433
	605-786	68.56	24.563	0.9927	
	786-819	4.58	208.341	0.9800	
A3	535-597	22.85	62.683	0.9920	45.318
	597-763	67.92	37.634	0.9945	
	763-794	3.62	150.110	0.9483	
B1	332-434	24.49	27.783	0.9950	20.882
	434-598	59.81	15.534	0.9987	
	598-651	11.83	40.465	0.9728	
B2	343-404	16.55	16.127	0.9874	15.943
	494-565	59.57	13.676	0.9988	
	565-617	13.43	38.176	0.9744	

Continued Table 2-35

Coal sample	Temperature range (℃)	Weight loss share (%)	Apparent activation energy of each stage E_i(kJ/mol)	Coefficient of association R	Average apparent activation energy E_m (kJ/mol)
B3	315-410	27.49	17.192	0.9960	14.973
	410-575	57.94	12.456	0.9953	
	575-610	7.56	40.079	0.9661	
C1	333-562	68.65	8.683	0.9969	10.563
	562-613	11.48	40.087	0.9504	
C2	313-555	68.53	7.487	0.9957	8.612
	555-613	12.58	27.671	0.9541	
C3	312-555	69.48	7.346	0.9968	7.952
	555-620	11.60	24.552	0.9768	

As can be seen from Table 2-35, the average apparent activation energy of pulverized coal decreases with the decrease of particle size, which is consistent with the research results of Jiang Xiumin and Sun Xuexin [6-7]. However, the average apparent activation energy of pulverized coal with different metamorphic degrees decreases with the decrease of particle size. The average apparent activation energy of pulverized coal decreases with the decrease of metamorphism. It can be seen from Table 2-35 that the average apparent activation energy of Baiyangshu anthracite with the highest degree of metamorphism decreases the most with the decrease of particle size. The average apparent activation energy of Jinkai lignite with the lowest metamorphism decreases the least with particle size. This is mainly because the composition of pulverized coal is different, containing volatile, fixed carbon is different. Baiyangshu anthracite has the lowest volatile content and the highest fixed carbon content. After pulverized coal is refined, the heat transfer rate of pulverized coal particles becomes faster, so that small particles are easy to reach the ignition temperature and ignition, and the energy and activation energy required are smaller. The volatile content of Jinkai lignite is higher, and the ignition mode of pulverized coal is different from that of Baiyangshu anthracite. The combustion mode of pulverized coal is mainly caused by volatile analysis, ignition and ignition of fixed carbon, so the reduction of pulverized coal particles has little influence on its combustion. It can be concluded from the above analysis that, in the experimental range, refining particle size is helpful to improve the combustibility of pulverized coal with high metamorphism. For pulverized coal with low metamorphism, particle size reduction has little effect on pulverized coal combustion.

2.3.5 Effect of particle size on combustibility of mixed coal

Taking scheme 3 (80% Xuanhua coal +20% Baiyangshu coal) as an example, the combustion characteristics of mixed coal powder with different particle sizes are studied. The experimental data are shown in Table 2-36.

Table 2-36 Coal mixing scheme 3 Combustibility of different particle sizes   (%)

Particle size(mesh)	500°C	550°C	600°C	650°C	700°C	750°C	800°C	850°C	900°C
50-100	39.58	54.25	67.03	79.56	91.37	98.54	99.97	100	100
100-140	39.95	54.74	67.88	80.87	93.23	99.73	100	100	100
140-200	44.42	59.43	73.08	86.73	98.16	99.76	100	100	100
-200	51.44	66.96	81.3	94.74	99.82	100	100	100	100

After analyzing the experimental data, the results are shown in Fig. 2-23.

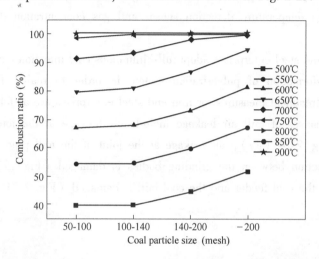

Fig. 2-23 Pulverized coal combustion curve of mixed coal

It can be seen from Fig. 2-23 that with the decrease of particle size, the combustion rate of pulverized coal mixture increases. The combustion rate of pulverized coal increases with the increase of combustion temperature. When the combustion temperature is higher than 700°C, the particle size of pulverized coal mixture in this experiment has little influence on the combustion effect of pulverized coal mixture. It is important to point out that in the actual production process of blast furnace, the high-speed jet of pulverized coal enters the raceway about area of tuyere and stays in the raceway about area for a very short time. Very short combustion time is the key to restrict the combustion rate of pulverized coal. Therefore, the particle size of pulverized coal should not be too large to avoid heat transfer and ignition problems caused by large particle size.

2.4 Full bituminous coal injection technology and industrial application

2.4.1 Safety assessment and rectification of blast furnace coal injection system

In view of the superior combustibility and low price of bituminous coal, the application of bituminous coal in iron and steel enterprises is becoming more and more popular, and its proportion in mixed coal powder is getting higher and higher. However, bituminous coal with high volatile content is flammable and explosive. In the actual production operation process, it is necessary to strictly carry out safety assessment and rectification of blast furnace pulverization and injection system to eliminate potential safety hazards and ensure production safety.

Due to the early construction of most iron and steel enterprises in my country, most of the equipment and detection systems in the current ironmaking workshops have some safety problems, which are specifically manifested in the following aspects:

(1) The mill runs for a long time and the airtightness is poor;

(2) The oxygen content in the dry preheating system is extremely high;

(3) The cloth bag is damaged and the air leakage is serious;

(4) The system temperature detection sensor and gas concentration detection sensor are damaged.

Therefore, iron and steel enterprises adopt full-bituminous coal injection technology to carry out more detailed transformation of pulverization system in order to realize full-bituminous coal injection. Common treatment measures for iron and steel enterprises are as follows.

(1) Common causes of mill air leakage include air leakage at the joint of the waste gas conveying pipe (Fig. 2-24 (a)), air leakage at the joint of the mill pipe (Fig. 2-24 (b)), and The soft connection between the grinding bodies is damaged (Fig. 2-24 (c)), the soft connection between the coal feeder and the coal mill is damaged (Fig. 2-24 (d)), and the air

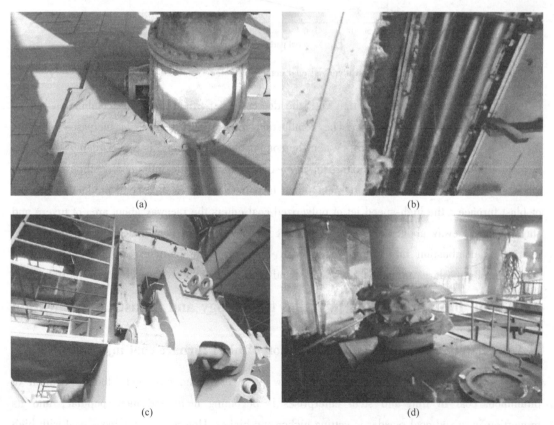

Fig. 2-24 Common causes of mill air leakage

leakage from the slag outlet of the mill (Fig. 2-25). The phenomenon of air leakage can be alleviated by adding rubber pads to the flange, replacing soft connections, and welding the air leakage of the metal box (Fig. 2-26). When the air leakage from the slag outlet of the grinding body is serious (because the slag outlet is always open), you should first contact the manufacturer to adjust and modify the structure of the air ring, or connect the exhaust gas to the slag outlet, enlarge the slag outlet and keep the slag outlet closed. At the same time, due to the low negative pressure of the slag discharge port, it is recommended to design the slag discharge port as a water-sealed type, insert the slag discharge port into the water tank, sealed with water and reduced oxygen in the mill by regular slag discharge.

Fig. 2-25 Air leakage from the slag outlet of the mill

Fig. 2-26 Repair air leakage by welding

(2) Find out the reason for the high oxygen content in the dry preheating system and deal with it in time. The main reasons for the excessively high oxygen content in the drying preheating system are: residual air from furnace replacement, air leakage from pipeline valves, residual oxygen from combustion, etc. The measures taken are as follows: reduce the excess air entering

the system due to the switching process of the hot blast stove by optimizing the operation, and reduce the oxygen substituted into the coal grinding system due to the operation of the equipment. Carefully check the airtightness of each part of the exhaust gas conveying pipeline, and reduce the oxygen entering the exhaust gas due to air leakage in the conveying system by adding rubber pads to the flanges at the joints. Through theoretical calculation or actual equipment detection, analyze whether the air excess coefficient of the heating furnace combustion gas is appropriate, and reduce the excess oxygen brought into the system due to excessive combustion air. Starting from the above aspects, reduce the oxygen entering the exhaust gas due to various reasons, thereby reducing the oxygen content at the entrance of the coal mill. As shown in Fig. 2-27, after system maintenance, the oxygen content in the mill and bag dropped significantly.

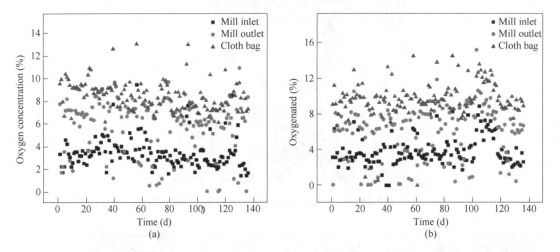

Fig. 2-27 Changes in oxygen content before and after system repair
(a) before repair; (b) after repair

(3) Reduce the temperature at the inlet and outlet of the mill. The inlet temperature of the mill can be controlled by adjusting the temperature of the drying gas, that is, by adjusting the ratio of the exhaust gas of the hot blast stove to the gas of the combustion furnace to control the temperature of the dry gas. If the inlet temperature of the mill is improved, the outlet temperature will be improved accordingly; at the same time, increasing the coal feeding will also reduce the outlet temperature of the mill.

2.4.2 Operating rules for blast furnace full injection of bituminous coal

2.4.2.1 General idea

Taking the current coal injection scheme of iron and steel enterprises as the benchmark, gradually adopt various optimized test schemes for production, and compare with the benchmark, step by step, and finally determine the optimal safe and economical coal injection scheme acceptable to the blast furnace and pulverizing system.

2.4.2.2 Benchmark industry test

According to the site situation, the current actual coal blending scheme has been continuously and stably produced for 14 days (the previous sampling test of the blended coal shows that the actual volatile content is close to 30%, and it is necessary to ensure that the volatile content of the blended coal is relatively stable. 25% of the preset level, and there will be no excessive volatile matter content exceeding the preset level), it is necessary to stabilize the amount of coal injection, raw material composition, equipment and furnace conditions as much as possible to ensure that there are no large fluctuations. Record relevant data on time during the benchmark test and use it as a comparison basis for subsequent optimization tests.

2.4.2.3 Formal industrial test

First select the optimized scheme with the composition of mixed coal closest to the base period for production test, and keep the same pulverized coal injection ratio, blast furnace raw fuel conditions and equipment conditions for continuous production for 14 days. According to the pulverizing system and the response of the blast furnace, it is determined whether to carry out the production test of the next optimal scheme. If there is no problem, then carry out the industrial test of the next optimization scheme, and the test period is still 14 days. By analogy, each test only changes the coal blending ratio, and keeps other conditions as consistent as possible until the industrial test is terminated for some reason or the tests of the four optimization schemes are successfully completed.

2.4.2.4 Determining the optimal coal blending ratio scheme

According to the industrial test in the previous stage, a comprehensive analysis of various ratio optimization schemes was carried out to determine a more economical and safer coal blending ratio scheme than the baseline, and to carry out continuous production for 14 days again with the determined scheme to investigate the production stability and effect reproducibility of the final optimized proportioning scheme.

2.4.2.5 Determine the reasonable injection ratio

After the optimal coal blending scheme is obtained, according to the production conditions, the injection ratio of the optimized coal blending scheme is gradually increased under the condition that other conditions are basically unchanged, and the test period for each change of an injection ratio is still 14 days. In this way, the economical and reasonable injection ratio range of the blast furnace is determined under the condition of applying the optimal coal blending ratio scheme for production.

2.4.2.6 Security requirements

A Temperature standard

(1) The inlet temperature of the coal mill is 280℃;

(2) The outlet temperature of the mill is less than 100℃;
(3) The temperature of the cloth bag is less than 95℃;
(4) The temperature of the finished coal bunker is less than 85℃.

B　The oxygen content standard of the system
(1) Oxygen content of exhaust gas from hot blast stove <3%;
(2) Oxygen content of coal mill inlet <6%;
(3) Oxygen content of coal mill outlet <6%;
(4) Oxygen content of bag box outlet <8%;
(5) Oxygen content of finished coal bunker <8%.

C　System CO concentration standard

Mill, cloth bag box, finished coal bunker $CO \leqslant 800 \times 10^{-6}$.

2.4.2.7　Other requirements

During the industrial test, every effort should be made to ensure the stability of single coal quality, the stability of blast furnace raw fuel (coke, sinter, pellets, etc.) quality, and the stability of production. The blast furnace thermal system, slagging system, and slag-discharging and iron-discharging system are adjusted normally, and the load is adjusted as appropriate according to the specific furnace conditions.

2.4.2.8　Industrial test content

During the entire period of the industrial test, various contents of the following processes were tested:

(1) Milling process. The injection mixed coal is sampled once a day for particle size analysis and proximate analysis.

(2) Blast furnace process. The person in charge of blast furnace production regularly records and counts the operating parameters and indicators of the blast furnace during the test period every day, and the ironworks is responsible for compiling the daily report of the industrial test.

2.4.3　Industrial application practice of blast furnace full injection of bituminous coal

Beijing University of Science and Technology and Tangshan Xinbaotai Co., Ltd. jointly successfully realized the full bituminous coal injection of the blast furnace in 2012. The blast furnace is safe and stable during the injection process and has achieved good economic benefits.

In the early stage, the research team of Beijing University of Science and Technology repeatedly studied the pulverized coal of iron and steel enterprises, and selected the bituminous coal of the power plant and the bituminous coal of the general plant as the main coal for the full injection of bituminous coal. Investigate the cause of air leakage, improve the equipment, and finally choose the No. 3 new medium-speed mill as the main mill for the bituminous coal injection experiment. During the industrial test, it is only used for the No. 2, 450m³ blast furnace.

Due to the special reason of Xinbaotai sintering, it is required to use sintered ore, pellet ore and lump ore as stable as possible. However, the instability of the charge during the industrial test still had a certain impact on this experiment.

In the course of the experiment, due to many reasons such as the market and inventory, firstly the ratio of lump ore was greatly increased, and the ratio of furnace charge structure was increased from 20% to about 30%, and then the auxiliary coke was replaced with Lu'an coke, which directly affected the strength. It has had an impact and is one of the main reasons for the increase in fuel ratio in the latter part of the industrial test.

The industrial test officially started on October 15th, 2012, and the bituminous coal ratio was increased from 50% to 100% bituminous coal injection. October 27, 2012 officially ended, a total of 13 days, Xinbaotai 100% bituminous coal injection was a complete success and achieved the expected goal.

2.4.3.1 Changes in particle size of injection coal during industrial tests

Since Xinbaotai Iron and Steel tried to produce 100% bituminous coal for the first time, the operators and on-site technicians were careful and carried out the grinding of coal powder completely according to the operating procedures. However, with the progress of the industrial test, the particle size of the pulverized coal prepared by the mill gradually becomes coarser, as shown in Table 2-37. When 100% bituminous coal is injected, the proportion of mixed coal smaller than 200 mesh (75μm) is 97.39% in the baseline period down to 45.29%. There are two main reasons for the coarsening of the mixed coal particle size: one is that the temperature of the mill is limited for the safety of pulverization, and the output of the coal mill is not enough, resulting in the coarser particle size of the coal powder; the other is that the bituminous coal can be ground. Therefore, with the increase of bituminous coal ratio, the grindability of the mixed raw coal becomes worse, and the particle size of the finished mixed coal gradually becomes coarser.

Table 2-37 Changes in coal particle size during industrial tests

Time	Name	<200 mesh (75μm) proportion (%)
October 17th	70%bituminous coal+30%anthracite	97.39
October 18th	70%bituminous coal+30%anthracite	50.95
October 20th	80%bituminous coal+20%anthracite	58.57
October 23th	80%bituminous coal+20%anthracite	71.23
October 26th	100%bituminous coal	45.29

2.4.3.2 Changes in composition of blast furnace dust during industrial test

The carbon content of blast furnace dust is mainly derived from coke powder and unburned coal powder, so the change of carbon content in blast furnace dust is usually used to evaluate the combustion rate of blast furnace coal. Blast furnace dedusting ash is divided into two types:

gravity ash and bag ash. Gravity dedusting ash is the dust removed from the blast furnace top flue gas after passing through the direct force dust collector. Usually, the particle size is relatively coarse, and the carbon-containing dust is mainly coke powder. Bag filter ash is the dust that is removed from the flue gas after direct force dust removal and then passed through the bag filter. Usually the particle size is finer, and the carbon-containing dust is mainly unburned pulverized coal. Therefore, the change of carbon content in bag dust is usually used to measure the combustion rate of pulverized coal in blast furnace. It can be seen from Table 2-38 that with the increase of bituminous coal content, the carbon content in bag dust removal ash does not change significantly, but the carbon content in gravity dust removal ash increases slightly. The analysis results show that when 100% bituminous coal is injected, the combustion rate of the blast furnace injection coal is basically unchanged, and a good combustion effect is maintained, but the degree of deterioration of the coke may be aggravated, resulting in an increase in the carbon content of the gravity ash.

Table 2-38 Carbon content in blast furnace dust during industrial tests

Date	Bituminous coal ratio (%)	Gravity dust (%)	Cloth bag ash contains carbon (%)
October 14th	50	17.50	24.43
October 16th	70	32.66	24.43
October 17th	70	22.37	25.31
October 18th	70	30.04	23.26
October 19th	80	20.31	28.64
October 20th	80	27.75	27.68
October 21th	80	23.19	23.79
October 22th	85	25.27	18.38
October 23th	85	26.99	25.13
October 24th	85	38.53	24.86
October 25th	100	35.84	22.78
October 26th	100	30.48	19.56

2.4.3.3 Changes of blast furnace operation index during industrial test

Fig. 2-28 shows the changes in the pressure difference of the blast furnace during the industrial test period. It can be seen that after increasing the ratio of lump ore in October 21h, the pressure difference fluctuated somewhat. On October 25h and October 26h, the 100% bituminous coal injection smelting pressure difference was normal, and the overall slight increase was not obvious on the 26th. Generally speaking, except for the fluctuation of pressure difference on October 21h, the pressure difference of the blast furnace was basically stable after adjustment in the rest of the hours.

2.4 Full bituminous coal injection technology and industrial application

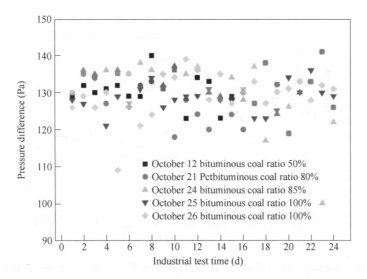

Fig. 2-28 Variation law of blast furnace pressure difference during industrial test

Fig. 2-29 shows the changes in the air permeability index of the blast furnace during the industrial test. It can be seen that the changes in the air permeability index are within the acceptable range of the blast furnace without abnormal fluctuations, and the full bituminous coal injection is generally better than other conditions. Therefore, from the point of view of pressure difference and air permeability index, there is no problem in the blanking of blast furnace.

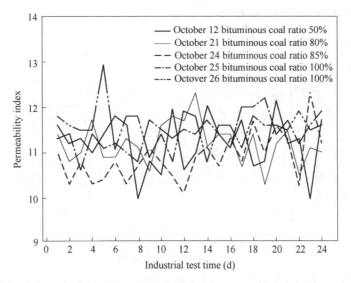

Fig. 2-29 Variation law of air permeability index of blast furnace during industrial test period

Fig. 2-30 shows the change of blast furnace utilization coefficient during the industrial test period. It can be seen that with the increase of bituminous coal ratio, the overall change of blast furnace utilization coefficient is not large, and fluctuates slightly between 4.02 and 4.28. Blast furnace production has not decreased, and the operation of the blast furnace is good.

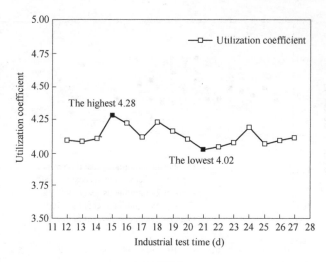

Fig. 2-30 Variation law of blast furnace utilization coefficient during industrial test period

Fig. 2-31 shows the change of silicon content in hot metal in blast furnace during industrial test. The silicon content of molten iron reflects the temperature of the molten iron. The higher the temperature of the molten iron, the more abundant the hearth temperature, and the higher the silicon content in the molten iron. The [Si] content was still high during the industrial test period, especially in the last 3 days of 100% bituminous coal injection. It shows that 100% bituminous coal injection does not cause the temperature of the blast furnace hearth to drop and affect the production of the blast furnace.

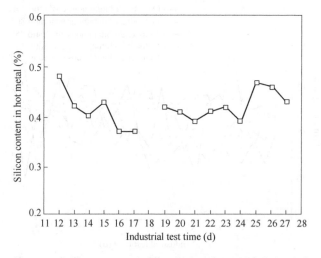

Fig. 2-31 Changes of silicon content in blast furnace hot metal during industrial tests

Fig. 2-32 shows the fuel consumption of the blast furnace during the industrial test period. It can be seen that the blast furnace coal ratio was stable in the early stage of the industrial test, but the blast furnace coal ratio continued to decrease in the later stage of the industrial test, and at the same time the blast furnace coke ratio also increased rapidly. In the case of technological indicators without considering economic benefits, it is best for Xinbaotai No. 2, 450m^3 blast

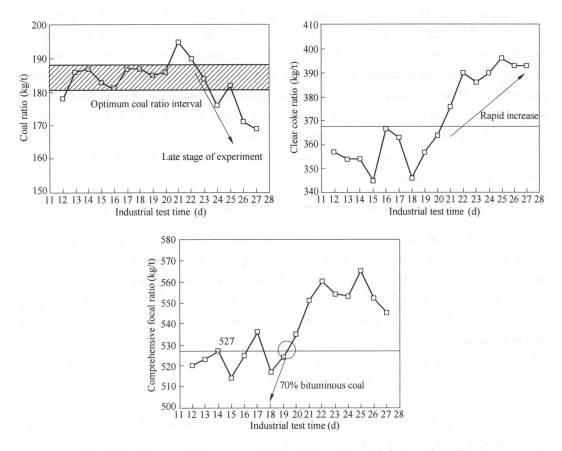

Fig. 2-32 Changes of silicon content in blast furnace hot metal during industrial tests

furnace to inject high volatile pulverized coal at around 185kg/t, not more than 190kg/t, and not lower than 181kg/t. The reduction of coal ratio will inevitably lead to the change of coke ratio. Since 20th, the coke ratio has been rising. The 20th coke ratio is 364kg/t, the coal ratio is 186kg/t, and the bituminous coal ratio is 70%-75%. Based on the current smelting situation, the best index of Xinbaotai blast furnace should be about 185kg/t coal ratio, about 360kg/t coke ratio, and the best bituminous coal ratio is 70%-75%. The index of Xinbaotai is the best. From an economic point of view, the current price ratio of bituminous coal and bituminous coal in the market does not apply to the 100% bituminous coal ratio, but the price difference between bituminous coal and bituminous coal will increase in the future, and Xinbaotai Iron and Steel Co., Ltd. is fully capable of full bituminous coal injection.

2.5 High proportion lignite injection technology and industrial application

By mixing and blending coal reasonably, the blast furnace can expand coal injection resources, reduce costs, and integrate the advantages of various coal types to achieve the best performance configuration of coal injection[8-9]. Some types of coal with wide range of coal sources, reasonable price, and poor performance indicators can also be properly applied when mixed coal injection is used. Among them, lignite with relatively low price has gradually attracted the attention of researchers [10-13]. Lignite is the coal with the lowest degree of coalification. It has high moisture content, low calorific value, and is easy to weather and spontaneously combust. The northeast and Inner Mongolia areas around Linggang are rich in lignite resources, which have certain geographical advantages, and the sulfur content of lignite is generally low. If it does not affect the performance of the coal injection process, adding lignite appropriately can effectively reduce the cost of coal injection. This section systematically analyzes the production practice of lignite injection in Linggang, aiming to provide reference for the expansion of blast furnace injection resources in other enterprises.

2.5.1 Current situation of coal injection in blast furnace of lingsteel

As a large state-owned iron and steel enterprise, Linggang has relatively advanced technology and complete equipment in China. The existing annual steel production capacity is 6 million tons, and the main equipment includes 5 blast furnaces, 6 converters, 5 continuous bar rolling lines, 1 high-speed wire rolling line, 1 medium-wide strip rolling line, and 9 welded steel pipe production lines. The process structure and product structure are reasonable and have strong market competitiveness.

The coal blending structure adopted by the blast furnace of the No. 2 ironmaking plant of Lingyang Iron and Steel Co., Ltd. is the mixed injection mode of "lean coal and bituminous coal". combustion effect. At present, under the conditions of 2% oxygen enrichment rate and 1200℃ air temperature, the blast furnace coal injection ratio is maintained at about 150kg/tHM, and the blast furnace is stable in forward motion.

There are 2 sets of medium-speed pulverizers in the No. 2 ironmaking and pulverizing workshop of Linggang. Currently, the 2,300m^3 blast furnace pulverized coal injection is provided by 2 sets of ZGM-95 medium-speed coal pulverizer vertical mills. The pulverizing capacity of the two mills is about 65t/h, the hourly coal injection volume of the 2,300m^3 blast furnace is about 38t/h in the reference period and the test period, respectively, and the pulverizing capacity of the coal mill can meet the injection requirements.

The coal mill currently used by Linggang controls the inlet temperature of the vertical mill to ≤ 300℃ and the outlet temperature of the vertical mill to 60-85℃ when grinding anthracite; controls the inlet temperature of the vertical mill to ≤ 260℃ when grinding bituminous coal (mixed coal)℃, the vertical mill outlet temperature is 60-75℃, the bag ash hopper temperature is ≤ 70℃, the pulverized coal bin temperature is ≤70℃, the front and rear shaft temperatures of the main exhaust fan are not higher than the ambient environment 40℃, all safety parameters are Control within a reasonable range[14].

2.5.2 Industrial experiment plan design

2.5.2.1 Industrial trial program

According to the site conditions, a period of time when the raw fuel conditions and operation of the blast furnace were relatively stable before the lignite industrial test was taken as the base period, and the production data during the base period when the furnace conditions fluctuated unsatisfactorily were ignored in the statistics and were not included in the scope of records.

Carry out the test according to the industrial test plan. During the test period, pay close attention to whether the pulverizing system is running normally and whether the blast furnace is running smoothly, and adjust the operation and industrial test arrangements in time according to the actual situation. Complete all coal blending schemes in the test scheme under the premise of safety and order. The test period is planned to be divided into 3 stages, and the specific test arrangement is as follows:

Stage 1: Coal blending structure was 10% lignite+90% electric cleaned coal, 6th-14th from January 2015;

Stage 2: Coal blending structure was 20% lignite+80% electrocleaned coal, from January 15 to January 20, 2015;

Stage 3: Coal blending structure is 30% lignite + 70% electrocleaned coal, 3rd-2nd from March 2015.

2.5.2.2 Industrial test guarantee

In order to ensure the smooth progress of the industrial test, before the start of the industrial test, a comprehensive analysis is made on the operational parameters control of the raw fuel management site and the safety control of the pulverizing and conveying system of Linggang No. 2 ironmaking, and combining with the actual situation of the site, the relevant improvement scheme is put forward.

A Strengthen safety management

Strictly monitor the temperature and oxygen content changes of the monitoring points of the pulverizing system and the conveying system, check the reliability of the emergency nitrogen flushing equipment, and ensure that the equipment is in a safe state during lignite injection. Establish specific system temperature control standards: the coal mill inlet temperature is less than 280℃, the mill outlet temperature is 70-100℃, the bag temperature is less than 95℃, and the coal bunker temperature is less than 85℃.

Oxygen content standard of the system: the oxygen content of the exhaust gas of the hot blast stove is less than 3%, the oxygen content of the inlet of the coal mill is less than 6%, the oxygen content of the outlet of the coal mill is less than 6%, the oxygen content of the outlet of the bag box is less than 8%, and the finished product Coal bunker oxygen content is less than 8%.

B Enhanced the quality stability of raw fuel.

During the industrial test period, ensure the stable supply of lignite as much as possible, and directly contact the production unit for purchase to ensure the stable quality of lignite. Strengthen the quality management of electro-cleaned coal, coke and other blast furnace raw materials (sinter, pellets, lump ore, etc.) to ensure stable quality of raw materials fed into the furnace, thereby stabilizing blast furnace production. During the industrial test, it is necessary to ensure the normal adjustment of the blast furnace heat system, slagging system, and slagging and ironing system to keep the furnace condition stable and smooth.

C Strengthen fuel storage management

To strengthen the management of the stockyard, the electric clean coal and lignite entering the plant are stacked separately, so as to avoid the influence of fuel mixing in the stockyard and the accuracy of retrieving materials. Make full use of the two coal storage bunkers in the pulverizing workshop of Linggang. During the test period, they are used to store electric clean coal and lignite respectively, so as to ensure that electric clean coal and lignite are fed separately. Calibrate the weighing of the coal blending belt to ensure the accuracy of the belt scale. By strengthening the stacking management and equipment maintenance of fuel delivered to the plant, the controllability and accuracy of the lignite blending ratio are guaranteed.

D Implementation of security measures

In order to prevent spontaneous combustion, the lignite stock is strictly controlled to be no more than 1,000t, and the storage period in the coal shed is no more than 5 days. During the test spraying period, the calibration is carried out once every 3 days, the process scale is accurate, the coal type and the coal bunker are installed with signs, and the boundary between the lignite and the coal piles on both sides is 2m. During the test spraying period, the bituminous coal canopy was used on average twice per shift (due to the large amount of coal slime and strong cohesion), but the interlocking system is easy to use and can be shut down in time. Strictly implement the daily inspection system to ensure that there is no dust or dust in the workshop, corridors, and equipment. Pay close attention to the composition of lignite, and carry out vehicle inspection on lignite entering the factory. Only after the test results come out, can it be used for production to guide a reasonable proportion.

2.5.3 Industrial experiment results analysis

2.5.3.1 Comparative analysis of blast furnace operation index

Table 2-39 shows the comparative analysis of blast furnace operating parameters during industrial experiments. It can be seen from the table that during the industrial experiment period, compared with the reference period, the air temperature was maintained at about 1,210℃, the oxygen enrichment was at 2.9%-3.1%, the furnace grade was at about 56.5%, and the air volume increased slightly. It shows that the air supply system and furnace grade are basically unchanged during the industrial experiment, which provides a good basic condition for the analysis of the influence of lignite injection on the blast furnace. Judging from the changes of top temperature, penetration index and theoretical combustion temperature, after lignite is used for blast furnace injection, the blast furnace top temperature slightly increases and the theoretical combustion temperature decreases slightly, but the penetration index remains basically unchanged. The main reason for the increase of blast furnace top temperature is that with the increase of lignite injection, the volatile content of pulverized coal increases, and at the same time, the wind at the tuyere increases, which leads to the increase of gas in the furnace, and the temperature of the furnace top rises accordingly. The reduction of theoretical combustion temperature is mainly due to the increase of lignite injection ratio, which leads to the increase of decomposition heat in front of the tuyeres, which reduces the temperature of combustion focus and the theoretical combustion temperature. Judging from the change of top temperature and physical combustion, it is completely acceptable to inject 30% lignite into the blast furnace.

Table 2-39 Changes of blast furnace operating parameters in each stage of baseline period and experimental period

Project	Utiliuzation coefficient (t/(m³·d))	Air volume (m³/s)	Wind temperature (℃)	Oxygen enrichment rate (%)	Taste in the oven (%)	Permeability index (m³/(min·kPa))	Top temperature (℃)	Secondary combustion (℃)
Base period	2.39	4412	1208	2.99	56.66	26.49	120	2300
Stage one	2.42	4415	1218	3.15	56.73	26.03	130	2309
Stage two	2.39	4436	1217	2.92	56.50	27.03	129	2301
Stage three	2.39	4476	1216	2.94	56.44	26.51	125	2274

2.5.3.2 Analysis of pulverized coal utilization

In order to investigate the utilization rate of pulverized coal at blast furnace tuyere during lignite injection, the carbon content of gravity dust in different stages was statistically analyzed on site, as shown in Fig. 2-33. It can be seen that with the increase of the ratio of lignite in the blast furnace PCI coal, the carbon content in the gravity dedusting ash decreases. The main reason is that lignite is a high volatile coal, and its combustion effect is better than that of bituminous coal and electric clean coal in the tuyere and raceway swirl area. The blending of lignite for blast furnace injection can effectively improve the combustion rate of tuyere pulverized coal and improve the utilization rate of pulverized coal.

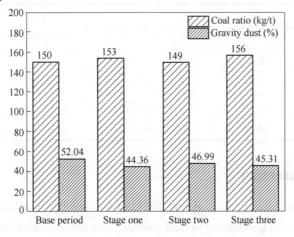

Fig. 2-33 Analysis results of carbon content in dedusting ash

2.5.3.3 Economic analysis of lignite injection

The purpose of lignite injection at Linggang is to give full play to the low price advantage of lignite, realize the replacement of bituminous coal with lignite for blast furnace injection, and reduce the fuel cost of blast furnace. Through statistical analysis, it is found that with the increase of lignite injection, the coal ratio has increased compared with the base period, the coke ratio has gradually decreased, and the fuel ratio has also gradually decreased, as shown in Table 2-40. It shows that the pulverized coal utilization rate of blast furnace tuyere increases after lignite injection at Lingyang Iron and Steel Co., Ltd., and the fuel consumption gradually decreases, which helps to greatly reduce the fuel cost of blast furnace.

Table 2-40 Blast furnace fuel consumption during baseline period and industrial test period

Project	Coal ratio(kg/tHM)	Focal ratio	Fuel ratio
Base period	150	395	545
Stage one	153	391	544
Stage two	149	389	538
Stage three	156	391	546

Table 2-41 shows the changes in fuel costs per ton of iron for blast furnaces during the base period and the industrial test period. It can be seen from the table that when lignite is used instead of bituminous coal for blast furnace injection, the cost of coal blending and coke is significantly reduced compared with the base period. With the addition of lignite increased to 30%, the cost of blast furnace fuel dropped from 624.91 yuan/t iron in the base period to 530.07 yuan/t iron. Excluding the influence of the drop in coke purchase price, the fuel cost per ton of iron decreased by about 7.81 yuan.

Table 2-41 Changes in fuel cost per ton of iron for blast furnaces during the baseline period and the industrial test period

Project	Cost of mixing coal (yuan)	Coke cost (yuan)	Power consumption (yuan)	Fuel cost (yuan)
Base period	90.1	531.81	3	624.91
Stage one	85.48	481.87	3	570.35
Stage two	82.72	476.23	3	558.95
Stage three	82.29	444.78	3	530.07

2.6 Chapter summary

(1) The selection and evaluation of coal for blast furnace injection should not only consider the proximate analysis and grindability of coal powder, but also consider calorific value, ultimate analysis, combustibility, reactivity, ash melting point, transportation performance, ignition

point, explosiveness, etc. Among them, the combustibility and effective calorific value of coal should be considered as key factors. Basic experimental studies have shown that low-rank coals (bituminous coal, lignite) have high combustibility and reactivity, which can help improve the combustion rate of pulverized coal. However, low-rank coal contains high volatile matter and strong explosiveness, so it is necessary to pay attention to the safety of pulverization and injection during use. In addition, the grindability of high-volatile bituminous coal and lignite is much worse than that of traditional PCI coal. When controlling the particle size of pulverized coal, it is necessary to comprehensively consider the pulverization capacity of the mill and the combustibility of pulverized coal.

(2) Tangshan Jianlong Xinbaotai Iron and Steel Co., Ltd. has mastered the safe operating procedures for pulverizing 100% bituminous coal injection and blast furnace injection technology. During the industrial test, the No. 2,450m^3 blast furnace ran smoothly, even under unfavorable conditions such as being forced to change the structure of the furnace material on a large scale, there was no fluctuation in furnace conditions. Based on the current smelting situation, the best index of Xinbaotai blast furnace should be around 185kg/t coal ratio, about 360kg/t coke ratio, and 70%-75% bituminous coal ratio.

(3) Linggang has successfully realized lignite instead of bituminous coal and electric clean coal mixed with thousand blast furnace injection. The proportion of lignite is gradually increased from 10% to 30%, and the blast furnace continues to operate stably until the successful completion of the industrial test. Through the analysis of the fuel cost during the industrial test period, it was found that when the addition ratio of lignite was 30%, the fuel cost per ton of iron in the blast furnace could be reduced by about 7.81yuan, lignite injection in blast furnace has a remarkable effect on reducing the cost of hot metal.

References

[1] Su Tianxiong. On the distribution and utilization of low-rank coal resources in China [J]. Guangdong Chemical Industry, 2012 (6): 141-142.

[2] He Xinjie, Zhang Jianliang, Qi Chenglin, et al. Effect of catalyst on combustion characteristics of pulverized coal and its kinetics [J]. Iron and Steel, 2012, 47 (7): 74-79.

[3] Jiang Xiumin, Li Jubin, Qiu Jianrong, et al. Study on combustion characteristics of ultra-fine pulverized coal [J]. Journal of China Electrical Engineering, 2000, 20 (6): 71-78.

[4] Zhou J H, Ping C J, Yang W J, et al. Thermo-gravimetric research on dynamic combustion re-action parameters of blended coals [J]. Power Engineering, 2005, 25 (2): 208-301.

[5] Cumming J W. Reactivity assessment of coals via a weighted mean activation energy [J]. Fuel, 1984, 63 (10): 1436-1440.

[6] Gu Fei, Gao Bin, Li Jianhui, et al. Influence of particle size composition on pulverized coal combustion rate [C] // 2001 Proceedings of China Iron and Steel Annual Meeting, 2001: 261-264.

[7] Sun Xuexin, Chen Jianyuan. Physical and chemical basis of pulverized coal combustion [M]. Wuhan: Huazhong University of Science and Technology Press, 1991.

[8] Chen Chunyuan, Zhu Yuying. Study on the reasonable proportion of bituminous coal injection in the new coal

injection system of baotou steel [J]. Baotou Steel Science and Technology, 2008 (4): 10-12.
[9] Xu Wanren. Limiting factors and technical measures to improve the coal injection quantity of blast furnace [C] // National Symposium on Coal Injection Technology for Blast Furnace, China, Jiangxi, Jingdezhen, 2008.
[10] Limpach R, Zhang Jianzhong. Experiment of injecting lignite into blast furnace [J]. Ironmaking, 1984 (4): 69-72.
[11] Zhang Qiumin, Wencui Li, Guo Shucai, et al. Study on the preparation of blast furnace injection and medium calorific value gas from Zhalainuoer lignite (I) Yield and properties of gas and tar [J]. Coal Conversion, 1997 (3): 69-73.
[12] Zhou Xigong. Exploration of lignite injection [J]. Journal of Kunming Institute of Technology, 1985 (1): 47-56.
[13] Zhang Wei, Wang Zaiyi, Zhang Liguo, et al. Exploration and practice of comprehensive injection technology for blast furnace in Angang [C] // The 9th China Iron and Steel Annual Meeting, China, Beijing, 2013.
[14] Ma Xiaoyong, Li Liang, Zhang Jianliang, et al. Production practice of lignite injection in Ling Gang 2300m^3 blast furnace [J]. Ironmaking, 2017 (3): 44-47.

Chapter 3 Technology and Industrial Application of Semi-coke Injection into Blast Furnace

3.1 Overview of semi-coke

The semi-coke, the structure is block, the particle size is generally above 3mm, the color is light black, it is the fuel and reducing agent in industries such as iron alloy, fertilizer, electric stones and blast furnaces, and is also the raw materials for chemical products such as activated carbon.

The earliest use of semi-coke started from the folk. Some places that were lacking in charcoal was in the stove. The remaining coal is removed and used for furnace fire or burn in the box. Due to the advantages of less smoke, high temperature, and flame resistance, etc., it is favored by rural housewives. When it burns, the flame is blue, so it is called semi-coke[1].

Semi-coke Industry began in Shen Mu and surrounding counties in the 1990s. The high-quality coal is burned with bright fire and the solid carbonized products obtained after the stack are extinguished. At the end of 2008 banning Turkey, semi-coke began the first round of upgrading and entered the era of charcoal furnace smelting. Originally, Anshan Thermal Energy Research Institute developed small furnaces with an annual production capucity of 50,000 tons, 30,000 tons, etc. to produce semi-coke, and recycled the coal tar produced during the dry distillation process. The produced equipment, currently, has completed the second round of upgrading and developed into a large-scale existing of more than 600,000 tons. The production device for generations, comprehensive utilization of charcoal, oil, and gas, large-scale production, energy saving, and automatic operation. The standardization of environmental protection and environmental protection has become a rookie in the coal chemical industry. Semi-coke, the industry, is based on the original indigenous industry. The industry that wanders on the edge of the policy has been widely recognized by the society after continuous technological progress and publicity and promotion, integrated into the national industrial catalog.

Now semi-coke, also known as "coke breeze, carbocoal", is a high-quality reducing agent for the production of ferroalloys and calcium carbide, which has the characteristics of three high and four low: high fixed carbon, high chemical activity, high specific resistance, low ash content, low aluminum, low sulfur and low phosphorus. The biggest advantage when using semi-coke to produce ferroalloys and calcium carbide is energy saving and consumption reduction (reduction of unit electricity consumption rate is 9.65%), especially when semi-coke is used to produce ferrosilicon and silicon alloy, the aluminum content of the product can be reduced and increased. The output rate of high-quality products can be increased, semi-coke has gradually replaced metallurgical coke, widely used in the produce of calcium carbide, ferroalloy, ferrosilicon, carbonization. Semi-coke has become an irreplaceable carbon material. And semi-coke can be divided into large grain material, semi medium material, semi small material and semi-coke foam according to the particle size, as shown in Fig. 3-1.

    Semi-coke material   Semi-coke medium   Semi-coke small material

Fig. 3-1 Semi-coke products

At present, the main production areas of semi-coke in China are Shaanxi, Xinjiang, Inner Mongolia and Ningxia, and the annual output of semi-coke is expected to be about 90 million tons. If this part of the semi-coke resource is applied to the steel industry, it will help reduce the steel industry's dependence on coking fertilizer coal and anthracite coal, optimize resource allocation and reduce environmental pollution[2]. However, how to reasonably apply different specifications of semi-coke to blast furnace injection and ensure the efficient and stable progress of the blast furnace ironmaking process is a key issue that needs to be solved at present.

3.2 Metallurgical properties of semi-coke

3.2.1 Proximate analysis of semi-coke

Three grain grades (semi-coke powder, small block of semi-coke, medium block of semi-coke, etc.) of Wuzhou, Hengyuan and Xingyong series of semi-coke produced in Shenmu area of Shaanxi were used as research samples, and Yangquan anthracite coal commonly used in domestic blast furnace injection was selected as a comparison sample. Each granular grade product is directly screened by the finished semi-coke, or obtained by screening after crushing the semi-coke. According to the national standard *semi-coke product varieties and grades* (GB/T 25212—2010), semi-coke products are divided powder particle size of semi-coke is <6mm, the small particle size of semi-coke is 6-13mm, and the medium particle size of semi-coke is 13-25mm.

Table 3-1 shows the promixmate analysis results of semi-coke samples and Yangquan anthracite. As can be seen from Table 3-1, the ash content of the semi-coke samples varied greatly, ranging from 4.5%-14.5%, except for Wuzhou semi-coke powder, Xingyong semi-coke powder and Xingyong of the medium block, they all meet China's requirements for coal ash content ($w_A < 12\%$) for blast furnace injection, and are slightly lower than Yangquan anthracite. Among them, the content of Hengyuan and Wuzhou semi-coke ash decreased with the grain level of the product, but there is no such rule for Xingyong semi-coke. The ash content of Hengyuan semi-coke is low as a whole, and the ash content in Hengyuan has the least ash content, only 4.58%, and Hengyuan semi-coke powder with the highest ash content in this series is only 10.19%. Compared with Hengyuan, the ash content of Wuzhou and Xingyong semi-coke is relatively high. Among them, the ash content of Wuzhou semi-coke powder is the highest, which is 14.42%. The second is Xingyong and Xingyong semi-coke powder, with the ash content of 13.18% and 12.62%, while the Xingyong small block is low, which is 8.00%. The reason for the difference in ash content is not only related to the composition of the raw coal used, but also to the amount of ash loss with smoke during the production process. In the application of blast furnace injection, appropriate combinations should be made according to the actual situation and experimental analysis.

Table 3-1 Proximate analysis of different types of semi-coke   (%)

Name	A_{ad}	V_{ad}	FC_{ad}	M_{ad}
Hengyuan semi-coke powder	10.19	9.88	79.29	0.64
Hengyuan small block	8.14	7.90	83.32	0.64
Hengyuan middle block	4.54	7.29	87.45	0.72
Wuzhou semi-coke powder	14.42	11.05	74.03	0.50
Wuzhou small block	8.42	8.77	82.07	0.74
Wuzhou middle block	8.34	6.69	84.21	0.76
Xingyong semi-coke powder	12.62	8.39	78.42	0.57
Xingyong small block	8.00	10.28	81.04	0.68
Xingyong middle block	13.18	7.52	78.65	0.65
Yangquan anthracite	9.95	7.06	82.36	0.63

Semi-coke is a carbocoal product that has been dried at medium and low temperatures, and its volatile matter has been extensively separated during the production process. From the perspective of component content, the semi-coke is closer to that of anthracite. From Table 3-1, it can be seen that the volatile matter of the semi-coke sample ranges from 6.5% to 11.5%, the volatile content of all other semi-coke samples is higher than that of Yangquan anthracite, the samples except for the Wuzhou middle block, which is 6.69%. Overall, there is not much difference in average volatile matter content among the three series of semi-coke products, while the content differences within each series are more significant with different particle sizes. Hengyuan semi-coke and Wuzhou semi-coke show a trend of increasing of volatile matter content as the particle size of the product decreases. This is likely caused by the uneven heating of raw coal during the production process, where a large amount of volatile from raw coal releasing and coking at higher temperatures, thus aggregating into relatively sturdy large blocks. At lower temperatures, the release volatile of raw coal is limited, and it cannot form solid semi-coke, which is easily broken into small pieces or powder.

The large precipitation of volatile matter leads to a relative increase in fixed carbon content in semi-coke. It can be seen from Table 3-1 that the fixed carbon content of semi-coke is relatively high, especially the fixed carbon content of Hengyuan middle block reaches 87.45%. Among the samples, Wuzhou semi-coke powder has the lowest content of 74.03%, and the fixed carbon content of other semi-coke samples is about 80%, similar to that of Yangquan anthracite. The fixed carbon content of Hengyuan semi-coke is generally the highest, followed by Wuzhou semi-coke, and Xingyong semi-coke is the lowest. The fixed carbon content of Hengyuan and Wuzhou semi-coke increases with the decrease of product particle size.

3.2.2 Ultimate analysis of semi-coke

Semi-coke is obtained from raw coal through medium to low temperature dry anchoring. During the production process, organic matter and some minerals are thermally resolved, resulting in a decrease in S, H, and O elements, especially sulfur content, which is extremely beneficial for steel production. The sulfur content of coal powder injected into the blast furnace should be the same as that of coke, generally considered to be less than 0.6%.

It can be seen from Table 3-2 that the sulfur content of various series of semi-coke samples is low, which is 30%-50% of the sulfur content of Yangquan anthracite, and the highest sulfur content of Hengyuan semi-coke powder is only 0.34%. Overall, Xingyong semi-coke has the lowest sulfur content, with an average sulfur content of 0.21%. Among them, Xingyong middle block has the lowest sulfur content of 0.18%, while Hengyuan semi-coke and Wuzhou semi-coke have roughly the same sulfur content, with an average sulfur content of around 0.26%.

Table 3-2 Ultimate analysis of different types of semi-coke   (%)

Name	C_{ad}	H_{ad}	N_{ad}	$S_{t,ad}$	O_{ad}
Hengyuan semi-coke powder	80.87	2.35	0.82	0.34	4.79
Hengyuan small block	84.54	1.82	0.80	0.22	3.84
Hengyuan middle block	88.63	1.78	0.87	0.24	3.00
Wuzhou semi-coke powder	75.18	1.78	0.70	0.30	7.12
Wuzhou small block	82.98	1.66	0.82	0.25	5.13
Wuzhou middle block	85.03	1.36	0.74	0.24	3.53
Xingyong semi-coke powder	79.13	1.79	0.86	0.26	4.77
Xingyong small block	82.98	2.06	0.80	0.20	5.28
Xingyong middle block	79.54	1.66	0.82	0.18	3.97
Yangquan anthracite	83.29	3.31	1.12	0.62	6.66

It can be seen from Table 3-2 that only the H content of Heng Yuan semi-coke powder in the semi-coke sample reaches 2.35%, and the H content of other samples is less than or equal to 2%, about 1/2 times that of Yangquan anthracite; The C content is between 80% and 90%, which is close to that of Yangquan anthracite, except that Wuzhou semi-coke powder is slightly lower. Due to the fact that the increase in gas content in the furnace hearth is related to H/C ratio of the fuel, the lower the H/C ratio, the less gas generated. Therefore, the loss of a large amount of H during the production process is highly likely to lead to a decrease in the gas content and penetration force of the furnace hearth after the injection of semi-coke into the blast furnace, resulting in a decrease in the preheating temperature of the upper coke and a decrease in the heat brought into the furnace hearth. The central heat cannot be replenished, resulting in a decrease in the central temperature of the furnace hearth and a corresponding reduction in the combustion zone. However, the decrease in H/C ratio also reduces the consumption of decomposition heat by the fuel, increases the effective heat release of the fuel in the tuyere and raceway, increasing the theoretical combustion temperature, and reducing coke consumption.

3.2.3 Ash composition analysis

Taking Hengyuan semi-coke as an example, the experiment investigated the CaO, MgO, Al_2O_3. The content of major oxides such as SiO_2 was analyzed and the results are shown in Table 3-3. The CaO content in the ash of different particle sizes of Hengyuan semi-coke products varies greatly, with Hengyuan semi-coke powder being the most abundant, reaching 15.9%. Next is the Hengyuan middle block (11.9%), Hengyuan small block is the least, at 9.23%. However, there is a small difference in the content of MgO and Al_2O_3, with MgO content ranging from 1.1% to 1.4%, the Al_2O_3 content ranges from 15% to 17%. Usually, the content of SiO_2 in coal ash is relatively high, with almost all minerals containing SiO_2. However, the content in Hengyuan semi-coke ash is relatively high, and there are significant differences in different particle sizes, ranging from 35% to 50%.

Table 3-3 Composition of crystal ash in Hengyuan semi-coke sample (%)

Name	CaO	MgO	Al_2O_3	SiO_2
Hengyuan semi-coke powder	15.9	1.14	15.30	35.30
Hengyuan small block	9.23	1.26	15.60	49.40
Hengyuan middle block	11.90	1.40	16.90	41.80

From Table 3-4, it can be seen that the Zn content of various particle sizes of Hengyuan semi-coke products is below 0.01%, which is lower than the lower limit of the detection range. The Na content is generally low, one order of magnitude lower than the pulverized coal used for normal blast furnace injection. With the increase of particle size, the Na content increases, which is consistent with the change trend of fixed carbon content of Hengyuan semi-coke. The K content of Hengyuan semi-coke shows a decreasing trend with the increase of particle size. The K content of Hengyuan semi-coke powder is higher, exceeding 0.4%, while the K content of Hengyuan small blocks is close to that of commonly used blast furnace injection coal powder. The K content of Hengyuan middle blocks is lower, only 0.0096%. The K in coal generally exists in the form of silicates such as feldspar and mica. The release of volatile matter has little effect on the K content in coal, while the impact of ash content becomes the main factor. Therefore, the higher the ash content, the higher the K content, which is consistent with the trend of decreasing ash content with the increase of particle size of Hengyuan semi-coke. The use of raw materials with low Zn and alkali metal content can reduce the damage of the harmful element to the coke skeleton and the corrosion of the furnace wall, which is conducive to the smooth production and the longevity of the blast furnace.

Table 3-4 Content of alkali metals and Zn in Hengyuan semi-coke   (%)

Name	Hengyuan semi-coke powder	Hengyuan small block	Hengyuan middle block
K	0.0453	0.0279	0.0096
Na	0.0450	0.0480	0.0550
Zn	≤0.010	≤0.010	≤0.010

3.2.4 Ash melting point

It can be seen from Table 3-5 that the ash melting characteristic temperature of the semi-coke sample is far lower than that of Yangquan anthracite. The four characteristic temperature ranges are: Deformation Temperature (DT) 1,084-1,195℃, Softening Temperature (ST) 1,168-1,290℃, Hemispherical Temperature (HT) 1,173-1,313℃, and Flow Temperature (Ff) 1,200-1,357℃. There are significant differences in the same characteristic temperature between different particle sizes of various series of samples, but there is no certain pattern between the two. The softening temperature is widely used and important among the four characteristic temperatures, and is generally used as a reference for the selection of combustion equipment. There are particle sizes with lower Softening Temperatures in all series of samples, that is, the ST<1,200℃ of this particle size; The Flow Temperature (FT) is also below 1,300℃, except for Wuzhou semi-coke powder and Xingyong middle block. Injecting samples with lower ash melting temperature into the blast furnace is highly likely to cause problems such as coal lance blockage and reduced combustion efficiency.

Table 3-5 Melting characteristics temperature of semi-coke sample crystal ash

Sample number	Sample name	Melting characteristic temperature/(℃)			
		DT	ST	HT	FT
1	Hengyuan semi-coke powder	1,137	1,178	1,195	1,223
2	Hengyuan small block	1,125	1,251	1,279	1,297
3	Hengyuan middle block	1,135	1,168	1,173	1,200
4	Wuzhou semi-coke powder	1,084	1,235	1,272	1,357
5	Wuzhou small block	1,148	1,229	1,256	1,289
6	Wuzhou middle block	1,143	1,172	1,184	1,200
7	Xingyong semi-coke powder	1,173	1,189	1,195	1,201
8	Xingyong small block	1,175	1,192	1,197	1,204
9	Xingyong middle block	1,195	1,290	1,313	1,336
10	Yangquan anthracite	1,465	>1,500	>1,500	>1,500

As shown in Fig. 3-2, the variation trends of the four characteristic temperatures of the semi-coke samples are roughly the same, but there is also a situation where the characteristic temperature difference span is large, such as the Wuzhou semi-coke powder. Its Deformation Temperature (DT) is the lowest at 1084℃, and the temperature difference between it and the Softening Temperature (ST) is large. The difference between the Softening Temperature (ST) and the Hemispherical Temperature (HT) is reduced, while the Flow Temperature (FT) is the highest in the sample, reaching 1357℃. The Wuzhou small block, Hengyuan small block, and Xingyong medium block, like the Wuzhou semi-coke powder, have a large characteristic temperature range, which mainly reflects the Deformation Temperature (DT) and Softening Temperature (ST). The remaining semi-coke ash samples form a sharp contrast, with a smaller temperature difference between the four characteristic temperatures. The curve shown in Fig. 3-2 is relatively flat.

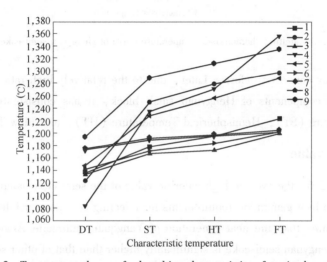

Fig. 3-2 Temperature change of ash melting characteristics of semi-coke samples

As shown in Fig. 3-3, there is no significant difference in Deformation Temperature (DT) among the three types of ash samples of Hengyuan semi-coke. The Deformation Temperature (DT) of Hengyuan small block is slightly lower than that of Hengyuan semi-coke powder and Hengyuan medium piece by about 10℃, while the difference in Softening Temperature (ST) is significantly increased. The Softening Temperature (ST) of Hengyuan small block is 1,251℃, which is about 65℃ higher than the other two types of ash samples. Afterwards, all characteristic temperatures show higher values, with a Flow Temperature (FT) of 1,297℃. Compared with Hengyuan medium block, the temperature advantage of ash melting characteristics of Hengyuan semi-coke powder gradually increases, but it is still significantly lower compared to Hengyuan small block.

The low temperature behavior of the Deformation Temperature (DT) of Hengyuan small block is likely due to the high content of SiO_2 in the ash. When the content of SiO_2 in ash ranges from 45% to 60%, SiO_2 is easily oxidized with other metals and non-metals forming a vitreous substance, which has an amorphous structure and exhibits a certain melting aid effect, So the

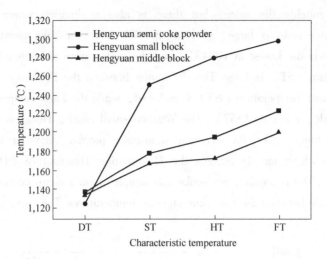

Fig. 3-3 Melting characteristic temperature curve of Hengyuan semi-coke ash

Hengyuan small gray cone first deforms. Later, due to the relatively high ratio of acid-base oxide content in the ash components of Hengyuan small block, it has consistently maintained high Softening Temperature (ST), Hemispherical Temperature (HT), and Flow Temperature (FT).

3.2.5 Heating value

As shown in Table 3-6, the average high calorific value of the semi-coke sample is 29.05MJ/kg, which can meet the heat generation requirements for injecting coal powder into the blast furnace, but overall, it is lower than the heat generation of Yangquan anthracite. Among them, the high calorific value of Hengyuan semi-coke is significantly higher than that of other series of semi-coke, and its high calorific value is above 29MJ/kg, the high calorific value of the Hengyuan middle block is the highest, reaching 31.51MJ/kg; The high calorific value of Wuzhou semi-coke powder is the lowest, only 26.33MJ/kg. According to the analysis method of effective heat, the effective heat released by burning Yangquan anthracite in the blast furnace tuyere is only 6.85MJ/kg, which is lower than the effective heat value of all other semi-coke except that it is higher than that of Wuzhou semi-coke powder.

Table 3-6 High calorific value of semi-coke

Name	High calorific value (MJ/kg)	Effective calorific value (MJ/kg)
Hengyuan semi-coke powder	29.20	7.59
Hengyuan small block	29.70	7.77
Hengyuan middle block	31.46	8.66
Wuzhou semi-coke powder	26.28	6.63
Wuzhou small block	29.44	8.08
Wuzhou middle block	30.00	8.38
Xingyong semi-coke powder	27.80	7.73

Continued Table 3-6

Name	High calorific value (MJ/kg)	Effective calorific value (MJ/kg)
Xingyong small block	29.29	7.73
Xingyong middle block	28.29	7.49
Yangquan anthracite	32.26	6.85

As shown in Fig. 3-4 (a), due to the dry anchoring of semi-coke at medium and low temperatures, the difference in volatile content among different semi-coke samples is small, only around 3%, so the impact of volatile content on heat generation is relatively small, In Fig. 3-4 (b), there is also an irregularity between volatile content and calorific value. The main factor affecting the calorific value of semi-coke is the difference in fixed carbon content. Generally speaking, the higher the fixed carbon content, the greater the calorific value of semi-coke.

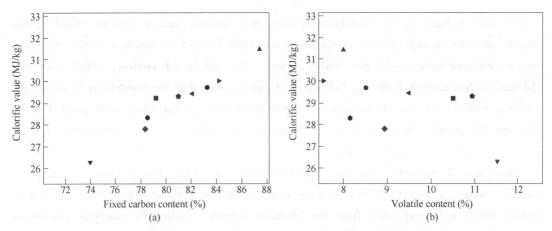

Fig. 3-4 Relationship between fixed carbon content and volatile matter content and calorific value of semi-coke sample
■—Hengyuansemi-coke powder; ◄—Hengyuan small block; ▲—Heng yuan middle block;
▼—Wuzhou semi-coke powder; ◄—Wuzhou small block; ►—Wuzhou middile block;
◆—Xingyong semi-coke powder; ●—Xingyong small block; ●—Xingyong middle block

3.2.6 Grindability

The Hardgrove grindability index of the semi-coke sample measured in the experiment is shown in Table 3-7. Compared with Yangquan anthracite, the Hardgrove grindability index of semi-coke sample is still lower. Among them, the grindability index of Hengyuan middle block is the highest, reaching 67.05. The grindability index of Hengyuan semi-coke powder is the lowest, only 54.97; Xingyong semi-coke powder is similar to it, with a value of 55.50. The remaining semi-coke samples are all around 60, with average grindability.

The grindability of coal is influenced by various factors: hardness, strength, toughness, and degree of dissociation are closely related to it. These properties are also influenced by the characteristics of coal metamorphism, coal petrographic composition, coal quality type, and mineral distribution. Semi-coke, on the other hand, is a semi-coke product obtained from medium and low temperature dry distillation of raw coal. During the production process, its hardness is improved compared to raw coal, and its internal structure and composition are more complex, making it difficult to discover its grindability from a single perspective.

Table 3-7 Hardgrove grindability index of semi-coke

Sample name	Grindability index	Sample name	Grindability index
Hengyuan semi-coke powder	54.97	Wuzhou middle block	62.90
Hengyuan small block	62.20	Xingyong semi-coke powder	55.50
Hengyuan middle block	67.05	Xingyong small block	61.51
Wuzhou semi-coke powder	64.98	Xingyong middle block	59.63
Wuzhou small block	58.74	Yangquan anthracite	68.44

As shown in Fig. 3-5, the correlation between ash content, volatile content, fixed carbon content, and the grindability of semi-coke is shown. From Fig. 3-5, it can be seen that there is a certain correlation between the grindability of semi-coke and its ash content, volatile content, and fixed carbon content. From Fig. 3-5 (a), it can be seen that the grindability of semi-coke decreases with the increase of ash content, mainly due to the fact that minerals are present in the ash, and the grindability of the minerals themselves is relatively poor. As the mineral content increases, the grindability of semi-coke also deteriorates. From Fig. 3-5 (b), it can be seen that the grindability of semi-coke is negatively correlated with its volatile content. From Fig. 3-5 (c), it can be seen that the grindability of semi-coke is positively correlated with the fixed carbon content. This can be explained from the following aspects: during the pyrolysis process of producing semi-coke, as the pyrolysis process becomes complete, the volatile content of semi-coke decreases and the fixed carbon content increases. During this process, semi-coke releases the volatile, the particle structure of semi-coke is severely damaged, and the strength of the carbon skeleton structure decreases. When the pyrolysis temperature increases, the side chains of semi-coke molecules break, and the generation and recombination of free radicals gradually increase the ordered carbon structure of semi-coke. The strength of the carbon structure will also increase, but it will reduce the grindability of semi-coke. Therefore, the pyrolysis temperature for producing semi-coke for blast furnace injection should not be too high.

3.2.7 Adhesion

The adhesion index of various series of semi-coke samples is 0, and there is no coking phenomenon between the mixture particles taken from the muffle furnace. This may be due to the large amount of organic matter precipitation during the pyrolysis process of raw coal at medium and low temperatures. During the process of injecting non cohesive semi-coke from the coal lance into the tuyere and raceway, it will not cause blockage of the coal lance and tuyere due to high-temperature coking.

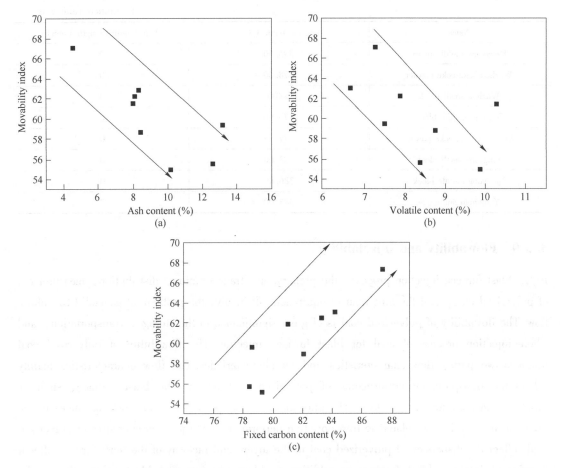

Fig. 3-5 Correlation between ash content, volatile content,
fixed carbon content and the grindability of semi-coke

3.2.8 Ignition point and explosivity

It can be seen from Table 3-8 that the ignition points of semi-coke samples are higher than 375℃, which is close to Yangquan anthracite. The highest is Wuzhou small block, and its ignition point is greater than 400℃, but the same series of Wuzhou semi-coke powder and Wuzhou middle block have lower ignition points. The ignition point of Hengyuan semi-coke powder is the highest as a whole, which is greater than 380℃. The ignition points of Xingyong semi-coke and Hengyuan semi-coke increased with the increase in product particle size, but the change range was small. The flame return length of all semi-coke samples is 0mm, that is, no explosion. In the production process, more bituminous coal can be added to ensure safety.

Table 3-8 The ignition point and explosiveness of semi-coke

Name	Ignition point (℃)	Flame return length (mm)
Hengyuan semi-coke powder	383.25	0
Hengyuan small block	385.85	0

Continued Table 3-8

Name	Ignition point (℃)	Flame return length (mm)
Hengyuan middle block	395.50	0
Wuzhou semi-coke powder	378.60	0
Wuzhou small block	>400	0
Wuzhou middle block	376.60	0
Xingyong semi-coke powder	377.55	0
Xingyong small block	383.00	0
Xingyong middle block	386.40	0
Yangquan anthracite	383.00	0

3.2.9 Flowability and injectability

In the blast furnace injection process, the preparation, transportation, distribution, measurement of pulverized coal, and the anti-wear of equipment all involve the problem of gas-solid two-phase flow. The flowability of pulverized coal is of great significance to the storage, transportation, and whole injection process of coal for blast furnace injection. The flowability of pulverized coal includes two parts: flow characteristics and jet characteristics. The flow characteristics mainly reflect the transportation performance of pulverized coal outside the blast furnace, such as pipeline transportation and stacking. The phenomenon of empty lances and blocking lances in the injection process is also related to transportation performance; the jet characteristics of pulverized coal reflect the dispersion of pulverized coal in the tuyere and raceway of the blast furnace after it is injected into the blast furnace in addition to blast pressure and blast volume. It can be considered that under the same external conditions, the greater the dispersion of pulverized coal in the tuyere and raceway, the higher the corresponding pulverized coal combustibility, the less the unburned coal powder, the higher the effective heat released by the corresponding pulverized coal, and the more obvious the purpose of replacing coke with coal.

It can be seen from Table 3-9 and Table 3-10 that the flow characteristic index of semi-coke is good, which is between 58.5 and 69.5. Except that the flow characteristic index of Wuzhou middle block and Xingyong semi-coke powder is low, which is 58.5, the flow characteristic index of other semi-coke samples is above 60, which is better than that of Yangquan anthracite. The jet characteristic index of semi-coke fluctuates greatly, ranging from 55.0 to 78.5, and there are also large differences among different series of products. Samples with good flow characteristics and jet characteristics, such as Wuzhou small block and Xingyong small block, can not only meet the transportation requirements of fuel in the pipeline, but also have good dispersion after being injected into the blast furnace, which is conducive to combustion. Compared with other semi-coke samples, Wuzhou middle block and Xingyong semi-coke powder are worse in flow characteristics and jet characteristics, which may affect smelting production in industrial applications.

Table 3-9　The flow characteristic index of semi-coke

Name	Repose angle (°)	Compression ratio	Plate angle (°)	Uniformity	Liquidity index
Hengyuan semi-coke powder	53.5	25.00	72	4.54	69.5
Hengyuan small block	50.0	20.70	69	4.89	64.5
Hengyuan middle block	50.0	25.30	74	5.25	62.5
Wuzhou semi-coke powder	45.0	26.00	75	4.90	62.0
Wuzhou small block	42.0	20.90	72	5.06	67.5
Wuzhou middle block	50.5	27.20	72	5.13	58.5
Xingyong semi-coke powder	50.5	29.10	68	5.19	58.5
Xingyong small block	41.5	17.79	82	5.28	63.5
Xingyong middle block	49.0	22.20	69	5.03	62.0
Yangquan anthracite	43.0	31.79	68	4.93	60.0

Table 3-10　The jet characteristic index of semi-coke

Name	Liquidity index	Collapse angle (°)	Difference angle (°)	Degree of dispersion (%)	Jet index
Hengyuan semi-coke powder	69.5	46.0	7.5	20	60.0
Hengyuan small block	64.5	46.0	4.0	31	58.0
Hengyuan middle block	62.5	44.5	5.5	37	65.0
Wuzhou semi-coke powder	62.0	40.0	5.0	33	62.0
Wuzhou small block	67.5	35.0	7.0	38	74.5
Wuzhou middle block	58.5	45.5	5.0	24	55.0
Xingyong semi-coke powder	58.5	44.0	6.5	24	58.3
Xingyong small block	63.5	30.0	11.5	48	78.5
Xingyong middle block	62.0	42.0	7.0	30	65.5
Yangquan anthracite	60.0	29.0	14.0	16	59.0

3.2.10　Combustibility

Under the experimental conditions, the changing trend of the weight loss curve of the combustion reaction of semi-coke is shown in Fig. 3-6. On the whole, the combustion weight loss of each semi-coke sample is close to Yangquan anthracite, and it is stronger than Yangquan anthracite in the early stage of combustion. With the advance of the combustion reaction and the increase of combustion temperature, the weight loss rate is gradually lower than Yangquan anthracite. The weight loss of semi-coke itself is quite different. The weight loss rate of Hengyuan semi-coke powder is higher than that of the other two, followed by Xingyong semi-coke powder, and Wuzhou semi-coke powder is relatively poor.

The combustion rate of semi-coke samples at characteristic temperatures is shown in Table 3-11.

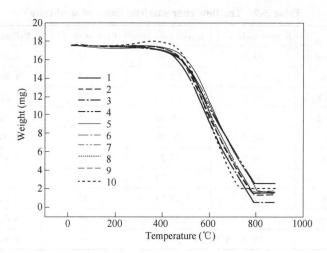

Fig. 3-6 Combustion weight loss curves of semi-coke and Yangquan anthracite samples

Table 3-11 Combustibility of semi-coke samples at characteristic temperature

Sample number	Coal types	Combustibility (%)		
		500℃	600℃	700℃
1	Hengyuan semi-coke powder	17.73	50.36	79.93
2	Hengyuan small block	15.35	46.06	76.93
3	Hengyuan middle block	14.58	47.56	77.12
4	Wuzhou semi-coke powder	13.47	44.39	74.21
5	Wuzhou small block	8.47	37.25	70.34
6	Wuzhou middle block	9.25	36.88	68.44
7	Xingyong semi-coke powder	11.98	40.58	71.52
8	Xingyong small block	13.81	43.11	75.42
9	Xingyong middle block	12.69	47.20	79.34
10	Yangquan anthracite	7.12	43.41	79.90

Fig. 3-7, Fig. 3-8 and Fig. 3-9 are shown as the columnar comparison of the combustion rate of 9 kinds of semi-coke and Yangquan anthracite at the characteristic temperature.

At 500℃, the combustion rate of each semi-coke sample varies greatly. The combustion rate of Hengyuan semi-coke is the highest, between 14.58% and 17.73%. The combustion rate of Xingyong semi-coke is between 11.98% and 13.81%. The burning rate of Wuzhou semi-coke is the lowest, between 8.47% and 13.47%. At this time, the combustion rate of semi-coke is higher than that of Yangquan anthracite. On the one hand, the pore structure produced by pyrolysis in the production process is conducive to combustion. On the other hand, because the residual volatile content in semi-coke is higher than that of Yangquan anthracite, it burns faster at low temperatures.

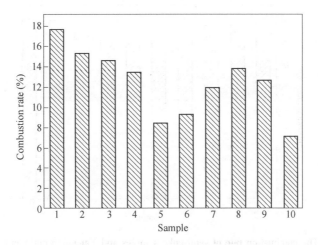

Fig. 3-7 The combustion rate of semi-coke samples and Yangquan anthracite at 500℃

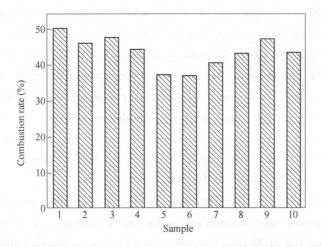

Fig. 3-8 The combustion rate of semi-coke samples and Yangquan anthracite at 600℃

At 600℃, the combustion rate gap of each semi-coke sample increnses. Except for some samples, the overall trend do not change, and the combustion rate of Yangquan anthracite increases significantly. The combustion rate of Hengyuan semi-coke is still higher than that of the other two semi-cokes, which is 46.06%-50.36%. The combustion rate of the Hengyuan middle block also increases from the third to the second at 500℃, and the combustion rate accelerates. Due to the different content of combustible materials in semi-coke samples, the lower the content of combustible materials in the same quality samples, the higher the combustion rate value may be, but the actual combustion quality may not be more. The Hengyuan middle block with more combustion quality can come back and catch up, indicating that its combustion performance is better. On the whole, the combustion rate of Xingyong semi-coke is still the second, and the order has not changed, which is 40.58%-47.20%. At this time, the burning rate of Wuzhou semi-coke is still the lowest among the semi-coke series, ranging from 36.88% to 44.39%.

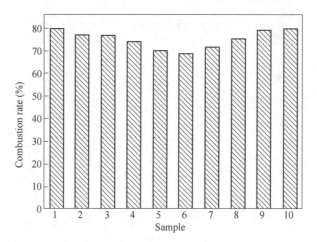

Fig. 3-9 The combustion rate of semi-coke samples and Yangquan anthracite at 700℃

At 700℃, the order of the burning rate of each semi-coke sample remains unchanged, but the gap between the burning rates is slightly reduced. It may be that since the combustible material has burned 70%-80% at this time, the combustion rate of each semi-coke sample has reached the limit, the combustion rate decreases rapidly with the combustion, and the combustion reaction process is close to the end, especially the combustion rate of the samples increases rapidly in the initial stage. At this time, the combustion rate of Hengyuan semi-coke is still the highest, 74.21%-77.12%; Xingyong semi-coke is relatively stable and still ranks second, and its combustion rate is 71.52%-79.34%. The difference in the burning rate of Wuzhou semi-coke is slightly smaller, between 68.44% and 74.21%.

On the whole, the order of combustion reactivity of semi-coke from strong to weak is Hengyuan semi-coke>Xingyong semi-coke>Wuzhou semi-coke. On the contrary, the stronger the combustion reactivity, the weaker the gasification reactivity with CO_2. This is closely related to the changes in the composition and structure of semi-coke during the production of medium and low-temperature carbonization.

3.2.11 Gasification characteristic

Under the experimental conditions, the comparative analysis of the weight loss curve of the semi-coke and CO_2 gasification reaction is shown in Fig. 3-10.

Table 3-12 shows the weight loss rate of the semi-coke sample and CO_2 gasification reaction at each characteristic temperature. From Table 3-12, it can be seen that under laboratory conditions, compared with Yangquan anthracite, semi-coke samples show strong reactivity with CO_2. All samples can react completely before 1150℃, and the difference in reaction performance between samples is also large. Overall, Wuzhou semi-coke has the strongest reactivity with CO_2. At 950℃, the weight loss rate of Wuzhou semi-coke samples with different particle sizes is between 9.83% and 10.69%, and the weight loss rate at each temperature is higher than that of Hengyuan semi-coke and Xingyong semi-coke. The Wuzhou semi-coke powder, small block, and

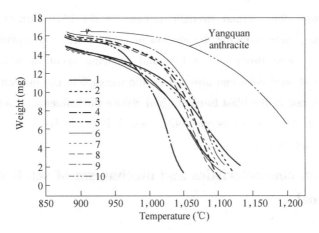

Fig. 3-10 The weight loss curves of semi-coke and Yangquan anthracite reacted with CO_2

middle block reacted completely at 1,093℃, 1,049℃, and 1,100℃ respectively, and the overall reaction temperature is lower than that of Hengyuan and Xingyong semi-coke. The reactivity of Xingyong semi-coke with CO_2 is slightly stronger than that of Hengyuan semi-coke. Although the weight loss rate of Xingyong semi-coke is lower than that of Hengyuan semi-coke at a lower temperature, the endpoint temperature of the complete reaction is lower than that of Hengyuan semi-coke. The complete endpoint temperature of the Xingyong semi-coke reaction is 1,100℃, 1,096℃, and 1,110℃ respectively according to the order of particle size from small. The complete endpoint temperature of the Xingyong semi-coke reaction is 1,132℃, 1,121℃, and 1,103℃ respectively according to the order of particle size to large.

Table 3-12 The weight loss rate of semi-coke and Yangquan anthracite reacted with CO_2

Number	Name	Weight loss rate (%)					
		950℃	1,000℃	1,050℃	1,100℃	1,150℃	1,200℃
1	Hengyuan semi-coke powder	10.06	21.72	43.21	79.99	100	—
2	Hengyuan small block	6.57	15.34	36.28	82.59	100	—
3	Hengyuan middle block	6.88	16.12	42.15	98.04	100	—
4	Wuzhou semi-coke powder	11.83	28.97	63.68	100	—	—
5	Wuzhou small block	9.83	37.05	100	—	—	—
6	Wuzhou middle block	10.69	27.90	59.06	96.93	100	—
7	Xingyong semi-coke powder	10.12	21.63	47.82	100	—	—
8	Xingyong small block	5.60	15.62	51.22	100	—	—
9	Xingyong middle block	5.43	13.06	37.33	94.58	100	—
10	Yangquan anthracite	1.26	4.13	10.00	21.16	40.73	67.54

The strong gasification reactivity of semi-coke with CO_2 can effectively protect coke in the middle and upper part of the furnace stack and reduce the gasification reaction of coke with CO_2.

This relatively improves the thermal strength of coke in the blast furnace and enhances the skeleton effect of coke, thus reducing the amount of coke used, and replacing coke with more pulverized coal in the blast furnace as a fuel and reducing agent. In addition, the excellent gasification reactivity of semi-coke can also reduce the incomplete combustion rate in the tuyere, improve its utilization rate in the blast furnace, and reduce the influence of unburned coal powder on the deterioration of blast furnace permeability, which is conducive to the smooth operation of blast furnace and reduce fuel consumption.

3.3 Combustion characteristics and mechanism of semi-coke mixed with injected coal

3.3.1 Experimental samples and methods

The experimental bituminous coal was taken from bituminous coal M for injection in an iron and steel plant, and semi-coke XJ was taken from a coal chemical industry Co., Ltd. The two fuels were dried in an oven at 40℃ for 3h to obtain air-dried base coal samples, crushed and screened, and 0.074-0.063mm pulverized coal was taken as experimental samples. The results of proximate analysis and ultimate analysis are shown in Table 3-13, and the results of ash composition analysis are shown in Table 3-14.

Table 3-13 Proximate analysis and ultimate analysis of coal samples   (%)

Coal	Industrial analysis				Elementary analysis				
	M_{ad}	A_{ad}	V_{ad}	FC_{ad}	C	H	O	N	S
Bituminous coal M	14.01	3.39	23.08	59.52	64.10	3.98	11.71	0.58	0.56
Semi-coke XJ	2.16	17.73	6.28	73.83	76.94	0.73	1.66	0.66	0.48

Table 3-14 Analysis of coal ash composition   (%)

Coal	CaO	SiO_2	Al_2O_3	MgO	Fe_2O_3	SO_3	TiO_2	Na_2O	K_2O	MnO	P_2O_5
Bituminous coal M	52.71	6.06	3.84	6.23	6.49	16.79	0.21	4.96	0.29	0.15	0.06
Semi-coke XJ	21.12	34.81	13.2	4.48	12.43	7.64	0.93	2.76	1.33	0.36	0.54

The mixed samples were mixed according to the bituminous coal M blending ratio of 20%, 40%, 60%, and 80%. The mixing method of pulverized coal is dry mixing milling for 10 min. The reference pulverized coal is also processed according to the above process to ensure that different coal samples have the same physical properties.

The combustion test was carried out by STA409 PC comprehensive thermal analyzer produced by Germany Necchi company. The diameter of the reactor is 60mm, the reaction atmosphere is air, and the inner diameter of the corundum crucible is 6mm. A total of 3.5mg sample was evenly tiled in the crucible, the initial temperature was room temperature, the heating rate was 15℃/min to 900℃, and the gas flow rate was 100mL/min.

3.3.2 TG/DTG curve analysis of combustion of single semi-coke and single bituminous coal

Fig. 3-11 shows the TG and DTG curves of XJ and M burning at a heating rate of 15℃/min. The TG and DTG curves of XJ and M are quite different during the heating process, indicating that the combustion process of the two is different. Generally, coal char combustion rises from room temperature to set temperature through the process of water evaporation, gradual precipitation of hydrocarbons in the form of volatiles, volatiles ignition, ignition of fixed carbon, or simultaneous ignition of fixed carbon and volatiles. From the curve, the various stages of pulverized coal combustion are crossed, and there is no obvious dividing line. The only difference between semi-coke XJ and bituminous coal M is that the DTG curve of bituminous coal M combustion stage is a single peak, while the DTG curve of semi-coke XJ has a small peak in the high-temperature range of 600-700℃, indicating that the semi-coke XJ has a slow combustion process at high temperature.

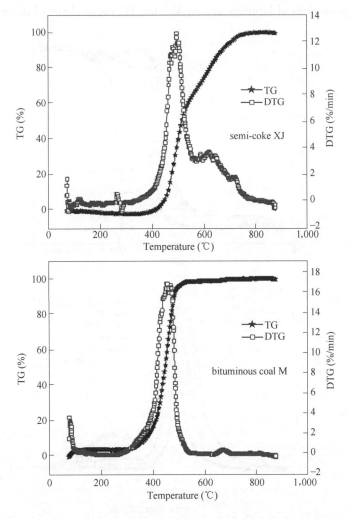

Fig. 3-11   TG/DTG curves of semi-coke XJ and bituminous coal M combustion alone

Compared with the shape of the DTG curve, the peak shape of the DTG curve of bituminous coal M is sharper and narrower than that of semi-coke XJ. This shows that bituminous coal M completed the violent combustion stage in a shorter time than semi-coke XJ. At the same time, the final inflection point temperature of the semi-coke XJ curve is later than the inflection point temperature of the bituminous coal M, indicating that the end temperature of the combustion process of the bituminous coal M is lower than the end temperature of the combustion process of the semi-coke XJ. From this point of view, the combustibility of bituminous coal M is better than that of semi-coke XJ.

3.3.3 TG/DTG curve analysis of mixed combustion of semi-coke and bituminous coal in different proportions

Fig. 3-12 shows the TG and DTG curves of bituminous coal mixed with 20%, 40%, 60%, and 80% bituminous coal M, 100% semi-coke XJ, and 100% bituminous coal M respectively. After adding different proportions of bituminous coal M, the TG curve shape of mixed coal is between

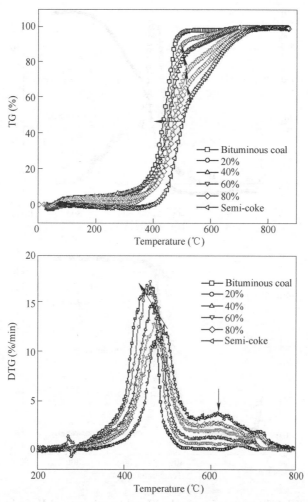

Fig. 3-12 TG/DTG curves of semi-coke XJ and bituminous coal M mixed combustion

bituminous coal M and semi-coke XJ, indicating that the combustion process of different mixed coals has a gradual change rule. With the increase of the bituminous coal M ratio, the TG curve of mixed coal gradually moves to the left, the slope becomes steeper and the curve gradually moves closer to the TG curve of 100% bituminous coal M. In addition, with the increase of the bituminous coal M ratio, the second inflection point of the mixed coal TG curve gradually moves to the upper left and finally disappears. This shows that with the increase of bituminous coal M ratio, the slow combustion process of the second stage of mixed coal gradually disappears.

Under the condition of different bituminous coal M ratios, the DTG curve peaks of mixed coal show a large single peak and a small peak. With the increase of bituminous coal M ratio, the DTG curve of mixed coal gradually moves to the left, and the peak shape gradually narrows. At the same time, with the increase of the bituminous coal M ratio, the small peak gradually becomes flat and finally disappears. From Fig. 3-12, it can be qualitatively analyzed that with the increase of bituminous coal M ratio, the flammability of blended coal gradually becomes better, and the combustion effect is gradually close to 100% bituminous coal M. This indicates that the combustion effect of bituminous coal M mixed with semi-coke XJ is greatly improved compared with the combustion of semi-coke XJ alone.

3.3.4 Combustion characteristic index analysis.

In order to quantitatively describe the combustion characteristics of pulverized coal, many scholars have proposed many combustion characteristic parameters based on the pulverized coal combustion process, including ignition temperature, flammability index, stable combustion index and comprehensive combustion characteristic index. In this section, when studying the combustion characteristics of mixed coal under different ratios of semi-coke, the commonly used ignition temperature (T_i, ℃), burnout temperature (T_f, ℃), flammability index (C, %/(min·℃^{2})), and comprehensive combustion characteristic index (S_N, %^{2}/(min^2·℃^{3})) are used for quantitative evaluation.

In this section, the TG-DTG method is used to determine the ignition temperature[3]. The flammability index C is used to represent the reaction ability of the mixed coal in the early stage of combustion. The larger the C value, the better the combustibility of the coal sample in the early stage.

$$C = W_{mean}/T_i^2$$

where, W_{mean} is the average burning rate, %/min.

Comprehensive combustion characteristics index S_N is used to evaluate the comprehensive combustion characteristics of coal samples. The comprehensive combustion characteristic index S_N represents the comprehensive combustion performance of coal, and the larger the S_N value, the better the combustion characteristics of coal.

$$S_N = \frac{W_{max} W_{mean}}{T_i^2 T_f}$$

where W_{max}——The maximum burning rate, %/min;

W_{mean}——The average burning rate, %/min;
T_i——The burnout temperature, ℃;
T_f——The ignition temperature, ℃.

Table 3-15 is the calculated combustion characteristic parameters of single coal and mixed coal. It can be seen from Table 3-15 that with the increase of the blending ratio of bituminous coal M, the ignition temperature of mixed coal decreases gradually, from 419℃ to 412℃; The burnout temperature of blended coal decreased gradually, from 752℃ to 697℃. This change rule is consistent with the conclusion obtained from the qualitative analysis of TG and DTG curves. Because the bituminous coal M contains high volatile and moisture, the volatile matter and water escape when heated, forming a good gas transmission channel, which is conducive to ignition and combustion. Semi-coke XJ has low volatile content and high ash content. Therefore, as the proportion of semi-coke M increases, the ash content in the mixed coal increases, which hinders the later stage of combustion, and the burnout temperature of the mixed coal gradually increases. From the point of view of flammability index C and comprehensive combustion characteristic index S_N, with the increase of the proportion of bituminous coal M, both of them gradually become larger. It shows that with the increase of the proportion of bituminous coal M, the combustion effect of blended coal becomes better. Plotting the flammability index and the bituminous coal M ratio, as shown in Fig. 3-13, it shows a certain linear relationship between the flammability of the blended coal and the bituminous coal M ratio. It shows that the flammability of blended coal mainly depends on the flammability and proportion of bituminous coal M, and the proportion of bituminous coal M can help improve the initial combustion effect of semi-coal XJ. From the relationship between the comprehensive combustion characteristic index and the bituminous coal M ratio, it can be seen that when the bituminous coal M ratio exceeds 20%, the comprehensive combustion characteristics of pulverized coal are almost linearly related to the bituminous coal ratio. This result shows that when the proportion of bituminous coal M is less than 20%, the comprehensive combustion effect of the mixed coal is determined by the combustibility of semi-coke XJ; when the proportion of bituminous coal M is greater than 20%, the comprehensive combustion effect of the mixed coal is determined by the bituminous coal M combustibility. This conclusion is important for guiding the use of semi-coke for blast furnace injecting.

Table 3-15 Characteristic parameters of blended coal combustion

Samples	T_i	T_f	T_{max}	W_{max}	W_{mean}	C (%/(min·℃²))	S_N (%²/(min²·℃³))
Semi-coke XJ 100%	438	764	496	12.47	4.23	2.20×10^{-5}	3.59×10^{-7}
Bituminous coal M 20%	419	752	478	11.48	3.93	2.24×10^{-5}	3.42×10^{-7}
Bituminous coal M 40%	414	727	473	12.89	3.95	2.30×10^{-5}	4.08×10^{-7}
Bituminous coal M 60%	413	693	469	14.8	4.15	2.43×10^{-5}	5.19×10^{-7}
Bituminous coal M 80%	412	697	465	17.14	4.11	2.42×10^{-5}	5.95×10^{-7}
Bituminous coal M 100%	398	673	456	16.85	3.98	2.51×10^{-5}	6.29×10^{-7}

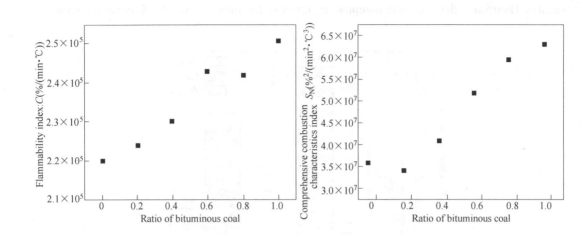

Fig. 3-13 Characteristic parameters of co-combustion of semi-coke XJ and bituminous coal M (C, S_N)

3.3.5 Mechanism analysis of factors affecting combustion characteristics

3.3.5.1 Composition analysis

From the perspective of proximate analysis, the bituminous coal M used in this study is characterized by high water content and volatile matter content, and the semi-coke XJ is characterized by high ash content, as shown in Fig. 3-14. During the combustion process of bituminous coal M, the volatilization of ash and the escape of volatile make the coal form more pore structures and accelerate its combustion. On the one hand, the ash content in pulverized coal can form a physical obstacle to the combustion of coal char and on the other hand, the ash content can play a catalytic role in the combustion of coal char. It is believed that the catalysis of K_2O, Na_2O, CaO, MgO and Fe_2O_3 in coal char can improve its combustion and gasification performance, and the presence of SiO_2 and Al_2O_3 can inhibit the combustion and gasification process of coal char. In order to determine the catalytic effect of minerals in the ashes of the two fuels on their combustion process in this section, the concept of catalytic index (A) was introduced[4-5]. The formula for calculating the catalytic index value is as follows:

$$A = W_A \times \frac{W_{Fe_3O_4} + W_{CaO} + W_{MgO} + W_{Na_2O} + W_{K_2O}}{W_{SiO_2} + W_{Al_2O_3}}$$

where, W_A is coal ash content, %; $W_{Fe_3O_4}$, W_{CaO}, W_{MgO}, W_{Na_2O}, W_{K_2O}, W_{SiO_2}, $W_{Al_2O_3}$ are percentage of Fe_3O_4, CaO, MgO, Na_2O, K_2O, SiO_2 and Al_2O_3 in ash, %, respectively.

The calculation results of the ash catalytic index of the two fuels are shown in Fig. 3-15. The calculation results show that the catalytic index of semi-coke XJ is 15.6, while that of bituminous coal M is 24.2. It shows that the catalytic effect of bituminous coal M ash is much greater than that of semi-coke XJ. At the same time, the ash content of bituminous coal M is much smaller than that of semi-coke XJ, and the physical hindrance to the combustion process is much smaller. Therefore, the combustion-supporting effect of bituminous coal M ash is more obvious.

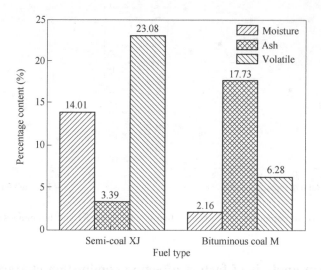

Fig. 3-14 Composition comparison of semi-coke XJ and bituminous coal M

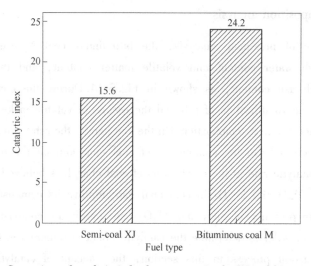

Fig. 3-15 Comparison of catalytic index between semi-coke XJ and bituminous coal M

3.3.5.2 Microstructural analysis

The combustibility of pulverized coal has a great relationship with its own morphology and structure. Fig. 3-16 (a) shows the morphology of semi-coke XJ under the scanning electron microscope, and Fig. 3-16 (b) shows the morphology of bituminous coal M under the scanning electron microscope. Semi-coke XJ is stone-like with no obvious pores, while bituminous M is flake-like and groove-like. The morphology characteristics of bituminous coal M are more conducive to the oxygen-coal contact and the precipitation of gas products during the combustion process, and the kinetic conditions of the combustion process are superior, which is one of the important factors that cause the combustion of bituminous coal M to be better than that of semi-coke XJ.

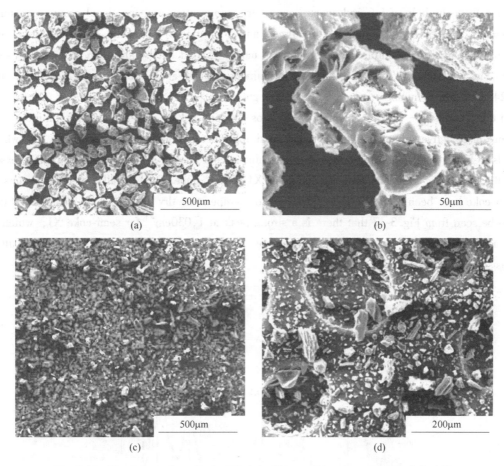

Fig. 3-16 Morphology of semi-coke XJ (a) (b) and bituminous coal M (c) (d)

3.3.5.3 Chemical Structure analysis

The main structure of coal is composed of carbon skeleton, and its periphery is various alkyl side chains and oxygen-containing functional group branch chains. Among them, the chemical properties of the carbon skeleton are relatively stable at low temperatures, while the group properties of the side chains and branched chains are relatively active. During the combustion process, the groups on the side chains and branch chains are easy to participate in chemical reactions, mainly the oxidation of active functional groups in coal and the precipitation of volatile. —OH is the most active organic functional group in coal. The oxidation reaction of —OH has a great damage to the molecular structure of coal, which can break the macromolecular chain in coal and decompose the carbon chain skeleton into smaller organic molecules. Therefore, its reaction process has a great influence on the combustion performance of pulverized coal. At the same time, relevant research also shows that the reactivity of functional groups such as —CH_3 and —CH_2 in coal is also strong, and these functional groups can be oxidized at low temperature.

Infrared spectroscopy is a widely used method for qualitative and quantitative analysis of organic matter. With the help of infrared spectroscopy, the composition and functional groups contained in single coal can be qualitatively analyzed and identified. At present, the qualitative and quantitative analysis of infrared spectroscopy is mostly carried out by absorbance. It can be seen from Fig. 3-17 that the peak positions of bituminous coal M at $3,400cm^{-1}$, $2,920cm^{-1}$, and $1,243cm^{-1}$ wave numbers, such as hydroxyl groups (—OH), fatty side chains (—CH$_3$), and ether bonds (—O—), are higher than those of semi-coke XJ. It shows that bituminous coal M has more active side chains and oxygen-containing functional groups than semi-coke XJ, which leads to better combustion performance of bituminous coal M than semi-coke XJ. The main reason for the low active group content of semi-coke XJ is that some active small molecular groups of semi-coke have been broken and escaped after low temperature dry distillation. In addition, it can also be seen from Fig. 3-17 that there is a strong peak at $1,030cm^{-1}$ for semi-coke XJ, which is the ash content peak, indicating that the ash content in semi-coke is very high, and this result is consistent with proximate analysis.

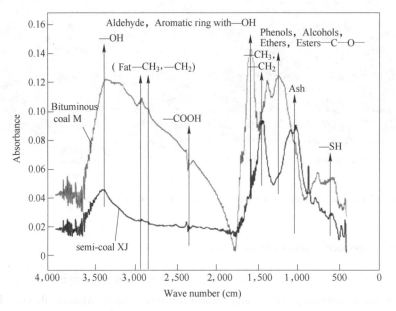

Fig. 3-17 Infrared absorption spectra of semi-coke XJ and bituminous coal M

3.4 Effect of semi-coke addition on metallurgical properties of mixed coal

3.4.1 Effect of semi-coke addition on safety performance of mixed coal

Using the aforementioned methods for determining the ignition point, explosiveness and ignition point of single coal powder, use the ignition point measuring device and long-tube type coal powder explosiveness testing equipment to measure the ignition point and explosiveness of semi-coal mixed with bituminous coal, and use a microcomputer automatic calorimeter to measure its calorific value, the measurement results are shown in Table 3-16.

Table 3-16 Ignition point, explosiveness and low calorific value of mixed coal of semi-coke and bituminous coal

Serial number	Experimental program	Ignition point (℃)	Explosive (mm)	Low calorifie value (J/g)
1	90% bituminous coal+10% semi-coke	311.45	320.5	23,720.18
2	80% bituminous coal+20% semi-coke	344.93	310	24,091.98
3	70% bituminous coal+30% semi-coke	347.73	256	24,547.28
4	60% bituminous coal+40% semi-coke	368.20	197	24,961.97
5	50% bituminous coal+50% semi-coke	376.00	96.5	25,374.38

The general trend of the ignition point change of blended coal is that the ignition point increases gradually with the decrease of bituminous coal content in the blended coal. According to the obtained single coal process performance, the ignition point of bituminous coal is obviously lower than that of semi-coke, and the ignition points of the five mixed coals obtained after the two single coals are mixed in proportion are basically between the respective ignition points of the two single coals, this is consistent with the volatile content of the blended coal being between the two single coals. The general trend of the explosive change of blended coal is that with the decrease of content, the explosiveness decreases obviously, and when the percentage of bituminous coal is greater than 70%, the explosiveness of blended coal is higher than that of any single coal, but they are all within the safe range (tempering length less than 400mm), indicating that the coal blending schemes all meet the safety requirements. The overall change trend of the calorific value of the blended coal is that with the increase of the bituminous coal content in the blended coal, the calorific value of the blended coal decreases, and the calorific value of the blended coal is between the two types of single coal, between 5,000 and 6,000kcal, calorific value is low.

3.4.2 Influence of adding semi-coke on the reactivity of mixed coal gasification

In the following, according to the coal blending scheme, the differential thermal analysis method is used to characterize the reaction performance of the five blended coals by measuring the weight loss rate at 1,000℃, 1,050℃ and 1,100℃. The measurement results are shown in Table 3-17.

Table 3-17 Reactivity of the blended coal of semi-coke and bituminous coal blends (%)

Serial number	Experimental program	1,000℃	1,050℃	1,100℃
1	50% bituminous coal+50% semi-coke	87.09	90.75	92.664
2	60% bituminous coal+40% semi-coke	87.82	95.65	98.47
3	70% bituminous coal+30% semi-coke	93.66	98.47	99.81
4	80% bituminous coal+20% semi-coke	94.73	99.24	99.90
5	90% bituminous coal+10% semi-coke	96.68	99.86	99.95

According to the measurement results, the reactivity relationship curves of five kinds of blended coal heated to 1,000℃, 1,050℃, and 1,100℃ in a CO_2 atmosphere and the histograms of combustion rates at each temperature point are drawn, as shown in Fig. 3-18 to Fig. 3-21.

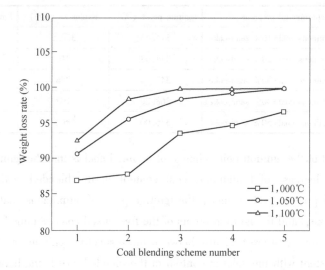

Fig. 3-18　Reaction characteristic curves of blended coal of semi-coke and bituminous coal

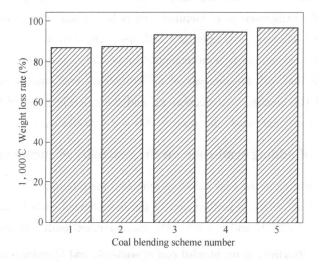

Fig. 3-19　Weight loss rate at 1,000℃ of blended semi-coke and bituminous coal

It can be seen from the figures that under the three temperature conditions of the experiment, the weight loss rate of the blended coal is above 80%, and the reaction characteristics are good. Under the same proportion, the reactivity of the blended coal increases with the increase of temperature. At relatively high temperatures, the reactivity increases slowly. Under the same temperature conditions, the reactivity of the blended coal increases with the increase of the proportion of bituminous coal in the blended coal. Among them, at 1,100℃, when the proportion of bituminous coal increases from 50% to 60%, the reactivity increases greatly; from

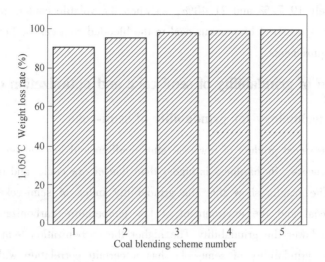

Fig. 3-20 Weight loss rate at 1,050℃ of blended semi-coke and bituminous coal

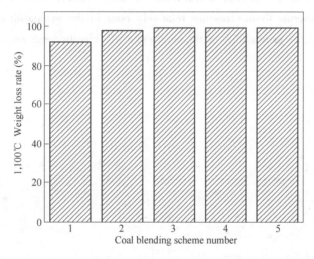

Fig. 3-21 Weight loss rate at 1,100℃ of blended semi-coke and bituminous coal

60% to 90% of the time, the reactivity is essentially unchanged. At 1,050℃ and 1,000℃, the reactivity increases greatly when the proportion of bituminous coal increases from 50% to 70%, and the change in reactivity is small when the proportion of 70% increases to 90%. Based on the above analysis, the reactivity is better when the proportion of bituminous coal in the blended coal is above 80%.

Among the five coal blending schemes based on the principle of coal blending, the coal blending schemes with a coal blending ratio of 80% or 90% of bituminous coal are within the safe range of explosiveness and low ignition point. More importantly, the combustibility and reactivity of these two blended coals are better in all coal blending schemes, that is, when these two coal blending schemes are selected, their utilization efficiency is higher under the same tuyere parameters. When the proportion of bituminous coal in the mixed coal is 80% and 90%, its volatile is respectively 19.72% and 21.40%, so when the volatile content of the mixed coal is around 20%, the proportion of bituminous coal in the blended coal can be flexibly controlled, to meet production requirements.

3.5 Regulation of grindability of semi-coke and optimization of injection size

3.5.1 Control technology for grindability of semi-coke

Semi-coke is the carbocoal product of raw coal after medium and low temperature carbonization, in the production process, its hardness is higher than that of raw coal, and its internal structure is more complex. The results show that the grindability index of semi-coke is related to the properties of raw coal and the environment of low temperature carbonization. The better the coalescency is, the lower the grindability. The higher the carbonization temperature, the lower the grindability. The grindability of semi-coke has a certain correlation with its ash content, volatile content and fixed carbon content. From Fig. 3-22 (a) can be seen that the grindability of semi-coke decreases with the increase of ash content, this is mainly because there are minerals in the ash, and the minerals themselves are relatively poor in the grindability. When the mineral content is increased, the grindability of the semi-coke will also become worse.

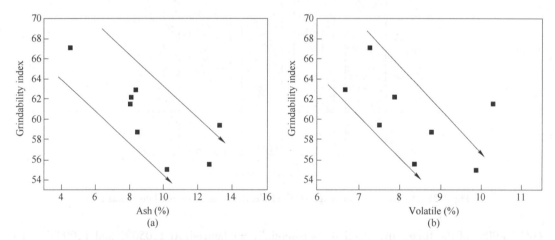

Fig. 3-22 Correlation between ash content, volatile and grindability of semi-coke

It can be seen from Fig. 3-22 (b) that the grindability of semi-coke is negatively correlated with its volatile content. In the pyrolysis process of producing semi-coke, the volatile content of semi-coke decreases and the fixed carbon content increases with the complete pyrolysis process, in this process, the volatile of semi-coke removes and the grain structure of the carbon is seriously damaged, and the strength of its carbon skeleton structure is reduced accordingly. However, when the pyrolysis temperature increases, the molecular side chain breaks, the generation and recombination of free radicals make the carbon structure order gradually improve, and the carbon structure strength also increases, reducing the grindability of the carbon. Fig. 3-23 shows the relationship between the grindability of semi-coke and the carbonization temperature of raw coal. With the increase of pyrolysis temperature, the grindability of semi-coke decreases. Therefore, in order to ensure good grindability of semi-coke in the injecting process, raw coal with low coalescence and low pyrolysis temperature are generally selected for production.

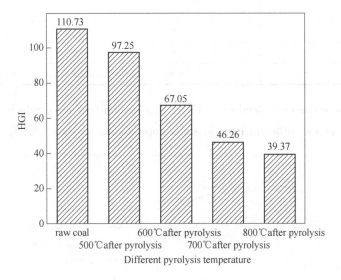

Fig. 3-23 Grindability of semi-coke at different pyrolysis temperatures

3.5.2 Optimization of mixing size of semi-coke and injection coal

In recent years, through the experiment of increasing coal ratio, it is found that relaxing pulverized coal particle size can save energy and reduce consumption, reduce pulverized coal manufacturing cost, and increase pulverized coal production capacity. However, too large pulverized coal particle size will affect the combustibility of pulverized coal, resulting in the increase of pulverized coal in the blast furnace raceway, and affect the lower permeability of the blast furnace. The national iron and steel enterprises according to the properties of their blast furnace pulverized coal find that the general requirements of pulverized coal particle size composition of -200 mesh ($-75\mu m$) proportion is not less than 70%, the best proportion is to reach more than 80%. At present, the mixed coal with 20% volatiles is obtained through the mixed of Altay bituminous coal and semi-coke. And the combustibility test of different particle size composition is carried out, bituminous and semi-coke are prepared into -200 mesh and 100-200 mesh (100 mesh = 150μm) particle sizes, then the two are of the same grain size pulverized coal is mixed proportionally to get two kinds of volatile content of 20%, but the particle size is -200 mesh and 100-200 mesh coal mixture respectively. Finally, the two kinds of coal are combined in different proportions, and seven experimental schemes are obtained when the proportion of pulverized coal in -200 mesh varies from 0% to 100%. Table 3-18 describes the specific schemes.

Table 3-18 Experimental scheme of coal mixing (%)

Number	−200 mesh ratio	100-200 mesh ratio
1	100	0
2	90	10
3	80	20
4	70	30
5	60	40
6	50	50
7	0	100

According to the above coal blending scheme, combustion tests are conducted to obtain the TG curves of mixed coal with different particle size compositions, as shown in Fig. 3-24.

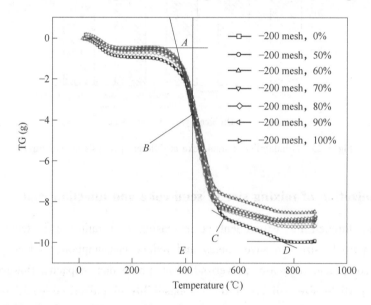

Fig. 3-24 Combustion TG curves of mixed coal with different particle size compositions

The temperature corresponding to the peak of the DTG curve is the temperature at which the weight loss rate is maximum. In Fig. 3-24, find the temperature point E corresponding to the maximum weight loss rate of a certain scheme on the horizontal axis, and make a straight line perpendicular to the horizontal axis through point E, intersecting with the TG curve at point B, where point B is the point at which the weight loss rate is maximum; Make a tangent at point B and intersect with the horizontal curve at point A at the beginning of the experiment. The temperature corresponding to point A is used as the starting temperature of coal powder combustion, which is called the ignition temperature; The inflection point temperature when the coal powder no longer loses weight, that is, the temperature at point D, is used as a sign of coal powder burnout, and this point temperature will be the burnout temperature; The point C is the approximate boundary point between volatile combustion and fixed carbon combustion, and it can be clearly seen that the combustion rate of coal powder before and after the point C varies greatly.

From the TG curve, it can be seen that overall, the TG curve shape of mixed coal with different particle size ratios is similar, that is, the particle size ratio has little impact on combustion performance, but the inflection point position, slope, and weight loss of each curve are slightly different. When the proportion of coal powder with smaller particle size in mixed coal is large, such as −200 mesh coal powder proportion accounting for 90% and 100%, the weight loss start temperature, burning temperature, and corresponding temperature at the maximum weight loss of coal powder are all lower

The combustion rate curves of mixed coal with different particle size ratios at 400℃, 500℃, 600℃, and 700℃ are shown in Fig. 3-25.

From Fig. 3-25 and Fig. 3-26, it can be concluded that the finer the particle size of blended coal at lower or higher temperatures, the better its combustibility; However, at higher temperatures, such as 600℃ and 700℃, there is a small difference in combustion rate between different particle size ratios of blended coal. Blended coal with the same particle size ratio has better combustion performance at higher temperatures, especially when the temperature increases from 400℃ to 500℃, the combustion rate significantly increases.

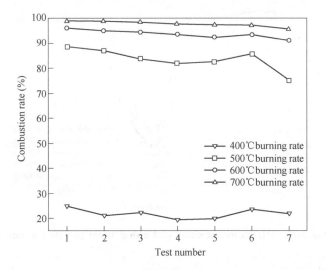

Fig. 3-25 Influence curve of particle size ratio on combustion performance

Fig. 3-27 show the DTG curves of blended coal under different particle size ratios. From the graph, it can be seen that overall, the finer the particle size of blended coal, the lower the temperature corresponding to the highest combustion rate during the combustion process, indicating better combustion performance During the process of changing the proportion of −200 mesh coal powder from 0% to 100%, the temperature decreases by nearly 30℃, and it is found from the upper and lower positions of the DTG curve peak, the maximum weight loss rate is higher when the proportion of −200 mesh coal powder is higher, indicating that its combustion effect is better.

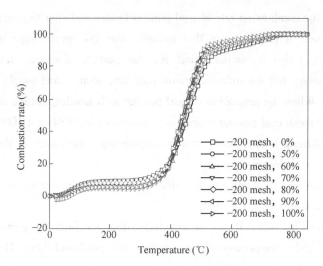

Fig. 3-26 Different particle sizes constitute the combustion rate curve of mixed coal

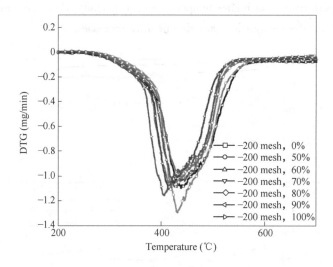

Fig. 3-27 DTG curves of mixed coal with different particle size ratios

Fig. 3-28 shows the ignition temperature, inflection point temperature, burnout temperature, and peak temperature corresponding to the DTG curve obtained by synthesizing TG and DTG curves with different particle size ratios. From Fig. 3-28, it can be seen that under all seven different particle size ratios, the particle size ratio has the smallest impact on the ignition temperature of coal powder, the largest impact on the burnout temperature, and the temperature corresponding to the inflection point temperature and the peak value of the DTG curve, i. e. the maximum weight loss rate, is between the above two. This may be because, from the perspective of coal petrology, coal powder is mainly divided into vitrinite, inertinite, and chitinite. In semi-coke, vitrinite is brittle and fragile, and it is very easy to damage vitrinite during coal grinding. Therefore, small particle size coal powder contains relatively more vitrinite, and vitrinite has better combustibility. Therefore, when the proportion of −200 mesh coal powder increases, coal has better combustion characteristics. From a physical perspective, when the particle size of coal powder is small, its specific surface area is relatively larger, which increases the chance of contact with oxidizing gases, improves heat and mass transfer conditions, and leads to more complete combustion. Due to the small difference in the proportion of −200 mesh coal powder among the 7 schemes, and considering the small amount of coal powder used in the experiment (10mg), poor mixing uniformity. The results obtained from the seven experimental schemes may not differ significantly, but the overall trend is that the larger the proportion of −200 mesh in the mixed coal, the better the combustion performance of the mixed coal.

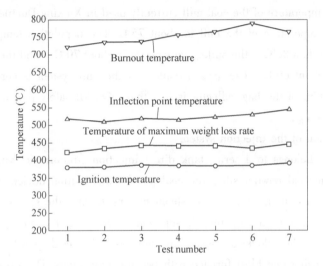

Fig. 3-28　Combustion characteristics of pulverized coal corresponding to changes in particle size ratio

3.6　Mixed injection technology and industrial application of semi-coke and pulverized coal

3.6.1　Current situation of coal injection into the blast furnace of Xinxing Ductile Iron Pipes Co., Ltd.

The industrial test of mixed injection of semi-coke and coal powder is carried out at Xinxing Ductile Iron Pipes Co., Ltd. Xinxing Ductile Iron Pipes Co., Ltd., as a large state-owned steel enterprise, has relatively advanced technical level and complete equipment level in China. There are two blast furnaces, 580m³ blast furnace and 420m³ blast furnace, obtaining three high-quality cast pipe production lines, complete equipment, and some processes that have reached the advanced level in China.

　　A　The situation of coal injection

The blast furnace in the Xinxing Ductile Iron Pipes iron adopts the injection mode of anthracite, which only injects one kind of pulverized coal, i.e. Luan anthracite, and the injection structure is relatively simple. At present, under the conditions of an oxygen enrichment rate of 3% and blast temperature maintain at 1190°C, the coal injection ratio of the blast furnace is maintained at around 150kg/tHM, and the blast furnace runs smoothly and stably.

B The situation of the milling system

There are two medium speed coal mills in the Xinxing Ductile Iron Pipes powder making workshop, one of which has been discontinued due to a malfunction. Currently, the coal powder injection for No. 1 and No. 3 blast furnaces is provided by another medium speed mill modeled E-1915. Its pulverizing capacity is around 50t/h, and the hourly coal injection rates of No. 1 and No. 3 blast furnaces during the reference and experimental periods are respectively at 11t/h and 13t/h, the powder capacity of the coal mill can meet the injection requirements.

The population temperature of the coal mill currently used in Xinxing Ductile Iron Pipes is 280-290°C, the outlet temperature of the mill is about 75°C. The population temperature of the bag dust collector is less than 80°C, the outlet temperature is about 70°C, and the temperature of the coal powder bin is about 60°C; The oxygen content of the mill system is controlled below 8%, and the oxygen content of the bag collector is less than 6%. All safety parameters are controlled within a reasonable range.

C The condition of the injection system

The injection system belongs to a series tank direct injection structure in Xin xing Dnctile Iron Pipes, including the coal powder silo, the coal powder distribution device under the silo, the injection tanks, the gas storage tanks, the steam heating tanks, the steam distribution drums, the injection pipelines, the valves, the air supplement devices, the coal powder distributors, the necking nozzles, the blockage measuring devices, the coal powder injection lances, etc. Direct injection using one blast furnace with two injection tanks. The storage capacity of the injection tank is about 22t, with nitrogen inside and compressed air as the carrier gas.

3.6.2 Determination of industrial experiment plan

3.6.2.1 Industrial test plan

According to the on-site situation, a period of time before the industrial test of injecting semi-coke into the blast furnace, during which the raw material conditions and operation conditions were relatively stable, was used as the reference period. Production data during the reference period when there were fluctuations in furnace conditions were ignored in statistics and were not included in the record range.

Conduct experiments according to the industrial test plan, closely monitor the normal operation of the pulverizing system and the smooth operation of the blast furnace during the experiment, and adjust the operation and industrial test arrangements in a timely manner according to the actual situation. Complete all coal blending plans in the test plan under the premise of safety and order. The experimental period is planned to be divided into three stages, and the specific experimental arrangement is as follows:

Stage 1: The coal blending structure is 10% semi-coke+90% Lu'an anthracite;

Stage 2: The coal blending structure is 20% semi-coke+80% Lu'an anthracite;
Stage 3: The coal blending structure is 25% semi-coke+75% Lu'an anthracite.

3.6.2.2 Industrial test guarantee

In order to ensure the smooth progress of the industrial test, a comprehensive analysis was conducted on the management of raw materials and fuels, on-site operation parameter control, and safety control of the powder conveying system for Xinxing Ductile Iron Pipes before the start of the industrial test. Based on the actual situation on site, relevant improvement plans were proposed.

A Strengthen safety management

Strictly monitor the temperature and oxygen content changes at the monitoring points of the pulverizing system and conveying system, check the reliability of the emergency nitrogen flushing equipment, and ensure that the equipment is in a safe state during the injection of semi-coke. Establish specific system temperature control standards: the mill population temperature is less than 280℃, the mill outlet temperature is between 70-100℃, the bag temperature is less than 95℃, and the coal bunker temperature is less than 85℃.

System oxygen content standards: the oxygen content in the exhaust gas of the hot blast stove is less than 3%, the oxygen content in the population of the coal mill is less than 6%, the oxygen content at the outlet of the coal mill is less than 6%, the oxygen content at the outlet of the bag box is less than 8%, and the oxygen content in the finished coal bunker is less than 8%

B Strengthen the stability of raw fuel quality

During the industrial trial period, try to ensure the stable supply of semi-coke, directly contact the production unit for procurement, and ensure the stable quality of semi-coke. Strengthen the quality management of Lu'an anthracite, coke and other raw materials (sinter, pellet, lump ore, etc.) of the blast furnace to ensure the stable quality of raw materials and fuels for the blast furnace, thus stabilizing the production of the blast furnace. During the industrial test period, it is necessary to ensure the normal adjustment of the blast furnace thermal system, slag making system, and slag tapping system to maintain stable and smooth furnace conditions.

C Strengthen the management of fuel stacking

Strengthen the management of the material yard, separate the Lu'an anthracite and semi-coke, and avoid the mixing of fuel in the material yard affecting the accuracy of material retrieval. Make full use of the two coal storage bunkers in Xinxing Ductile Iron Pipes' pulverizing workshop. During the test, they are respectively used to store semi-coke and anthracite to ensure that semi-coke and anthracite are separately fed. Correct the weighing of the coal blending belt to ensure the accuracy of the belt scale. By strengthening the stacking management of fuel and equipment maintenance at the factory, we ensure the controllability and accuracy of the proportion of semi-coke.

3.6.3 Comparative analysis of blast furnace smelting parameters

3.6.3.1 Comparison and analysis of injection mixed coal

Before the industrial test, Xinxing Ductile Iron Pipes separately injected Lu'an anthracite, the coal ratio was maintained at about 152kg/tHM, the coke ratio was about 404kg/tHM (including coke briquette), the fuel ratio was about 556kg/tHM, and the furnace condition was stable and smooth. Shenmu semi-coke is a kind of coal chemical product whose composition is close to that of bituminous coal, and its combustibility and reactivity are better than that of ordinary anthracite. The purpose of this industrial test is to mix Lu'an anthracite with semi-coke to reduce production costs while maintaining smooth furnace conditions. During the experiment, it is necessary to ensure the stable quality of coal powder injection. Table 3-19 shows the mean values of the main components of coal powder injection at each stage.

Table 3-19 Composition of coal injection at different stages (%)

Item	Moisture	Ash content	Volatile matter	S	−200 mesh ratio
Base period	1.28	10.39	12.31	0.38	74.27
Stage 1	1.23	10.83	12.30	0.35	73.55
Stage 2	1.36	11.13	12.56	0.35	74.56
Stage 3	1.32	11.14	12.78	0.34	74.45
Coke	5.64	12.88	1.03	0.76	1.03

From Table 3-19, it can be seen that compared to separately injecting Lu'an anthracite, the water content, ash content, and volatile lonternt of the mixed coal increase with the addition of 10%, 20%, and 25% of semi-coke, which is consistent with the previous theoretical analysis. The principle of coal blending in industrial experiments is to ensure that the moisture content of the mixed coal is within 1.5%, and the ash content of mixed coal is less than 12%. From the composition of mixed coal, it can be seen that the composition indicators of the mixed coal are still within the control range after adding 25% to the semi-coke. In addition, due to the low sulfur content in semi-coke, the sulfur content of mixed coal gradually decreases with the increase of semi-coke proportion, which is more advantageous for reducing desulfurization heat consumption.

The fluctuation of coal ash and sulfur content during the reference and experimental periods is shown in Fig. 3-29.

From Fig. 3-29, it can be seen that compared to the reference period, the sulfur content of mixed coal in each stage of the experimental period fluctuates more significantly, especially in stage one, but overall it shows a downward trend. Compared with the reference period, the ash content of mixed coal in Stage 1 has increased slightly, but the increase is relatively small. The ash content of mixed coal in Stage 2 and Stage 3 is basically the same as that in the later stage of Stage 1. Overall, the ash content of mixed coal in each stage of the experimental period has slightly increased compared to the reference period, but the increase is not significant and is all below 12%, which can meet the production requirements of the blast furnace.

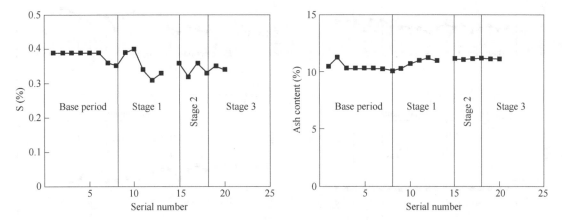

Fig. 3-29 Fluctuation of mixed coal during the reference and experimental periods

3.6.3.2 Comparative analysis of blast furnace operation indicators

The industrial experiment of injecting semi-coke aims to reduce fuel costs, but the stability and smooth operation of the blast furnace is the most fundamental prerequisite. If the condition of the blast furnace after injecting semi-coke is not at the top, or if there are problems with the smooth operation of the blast furnace after injecting a certain proportion of semi-coke, then reducing fuel costs can start from scratch. Therefore, it is necessary to observe and adjust the reaction of the blast furnace after injecting different proportions of semi-coke to ensure that the goal of reducing the cost of injecting fuel is achieved while the blast furnace is stable and smooth. Table 3-20 shows the average values of the main operating parameters of the blast furnace during the reference and experimental periods, and Fig. 3-30 shows the daily variation of each parameter.

Table 3-20 Changes in operating parameters of No. 3 blast furnace during the baseline and experimental periods

Item	Utilization coefficient ($t/(m^3 \cdot d)$)	Air volume (m^3/s)	Wind temperature (℃)	Oxygen enrichment rate (%)	Furnace taste (%)	Transparent finger ($m^3/(min \cdot kPa)$)	Top temperature (℃)	Theoretical combustion temperature (℃)
Base period	3.74	1,609	1,188	3.01	55.50	11.40	222	2,239
Stage 1	3.85	1,626	1,189	2.69	55.43	12.16	217	2,230
Stage 2	3.93	1,629	1,189	2.84	55.81	12.22	214	2,227
Stage 3	3.95	1,635	1,188	2.62	56.07	12.04	211	2,225

During the industrial test period, the blast furnace was stable and smooth, receiving 25% mixed injection of semi-coke. From Table 3-20 and Fig. 3-30, it can be seen that compared to the reference period, the charging grade, blast temperature, oxygen enrichment, and blast volume in each stage of the experiment are basically at the same level, while the blast temperature

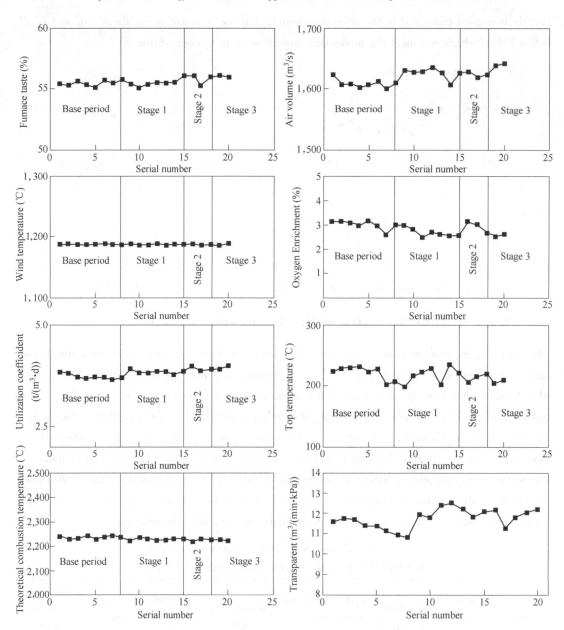

Fig. 3-30 Changes in operating parameters during the reference period and industrial test period of blast furnace No. 3

remains almost unchanged. The charging grade slightly increases, the blast volume slightly increases, and the oxygen enrichment rate slightly decreases. From the perspective of blast furnace production, the utilization coefficient slightly increased during the experimental period, mainly due to a slight increase in the grade of the iron ore, an increase in blast volume, and a stable operation of the blast furnace during the experimental period, resulting in an increase in production. From the smoothness of the blast furnace, the permeability index slightly increases and the theoretical combustion temperature slightly decreases.

The increase in permeability index indicates that the internal permeability of the blast furnace has improved, and the operating conditions of the blast furnace have improved. On the one hand, it is due to the increase in the primary combustion rate in the tuyere and raceway of the blast furnace after the mixed injection of semi-coke, the consumption rate of unburned coal powder has increased, and the permeability of the soft melt zone has gradually increased. On the other hand, the grade of the blast furnace has slightly increased, and the amount of iron slag per ton has decreased, which is beneficial for improving the permeability inside the blast furnace. The decrease in theoretical combustion temperature is mainly due to an increase in moisture and ash content in the mixed coal. In the tuyere and raceway, the moisture in the coal undergoes, and heat absorption of gasification and decomposition, as well as the heat consumption of ash into slag, causing a decrease in the theoretical combustion temperature. However, from the industrial test results, the theoretical combustion reduction is still greater than 2,200℃, which can meet the heat required for blast furnace production, and the thermal state of the furnace hearth is sufficient. From the values of furnace top temperature at each stage, it can be seen that the furnace top temperature during the experimental period has slightly decreased compared to the reference period, indicating an improvement in gas utilization efficiency.

3.6.3.3 Analysis of coal powder utilization efficiency

The carbon content in blast furnace dust can reflect the utilization efficiency of coal powder after increasing the amount of coal injected. During the industrial trial period, the quality of coke was slightly better than the reference period, and the quality of coal powder fluctuated significantly. The changes in carbon content of dust removal ash during industrial testing are shown in Table 3-21.

Table 3-21　Analysis results of carbon content in dust removal ash of No. 3 blast furnace

Item	Coal ratio (kg/tHM)	Gravitational ash content (%)
Base period	148	40.25
Stage 1	155	37.57
Stage 2	155	39.53
Stage 3	157	40.23

From Table 3-21, it can be seen that the coal ratio in the two stages of the experimental period has significantly increased, and compared to the reference period, the blast furnace's blast temperature level remains almost unchanged, while the oxygen enrichment level has decreased. In this case, the carbon content in the gravity ash still slightly decreases, indicating that injecting mixed coal powder of semi-coke and Lu'an coal is more easily utilized in the blast furnace, improving the utilization efficiency of coal powder in the blast furnace.

3.6.3.4 Economic analysis of semi-coke injection

The purpose of injecting semi-coke and Luan anthracite blended pulverized coal is to reduce the fuel cost. How the coke ratio and fuel ratio of the blast furnace change after injecting semi-coke, and how the fuel cost changes are the most important concerns of the iron and steel enterprises, which is also the core issue of this industrial test. Therefore, it is necessary to make statistics, analysis and summary of the changes in coke ratio, fuel ratio and fuel cost during the reference period and industrial test, and provide reasonable suggestions for the reasonable injection ratio of semi-coke. Table 3-22 shows the changes in coal ratio, coke ratio (including nut coke), and fuel ratio of No. 3 blast furnace during the reference period and industrial test period.

Table 3-22 Fuel consumption of No. 3 blast furnace during the reference period and industrial test period

Item	Coal ratio (kg/tHM)	Coke ratio (kg/tHM)	Fuel ratio (kg/tHM)
Base period	148	394	542
Stage 1	155	387	541
Stage 2	155	388	542
Stage 3	156	387	543

From Table 3-22, it can be seen that compared to the reference period, the coal ratio during the experimental period has increased, the coke ratio has decreased, and the fuel ratio is basically the same. After increasing the proportion of semi-coke injection, the change in fuel ratio is relatively small. It shows that the utilization efficiency of pulverized coal increases after the mixed injection of semi-coke and Lu'an anthracite. The replacement ratio of coal and coke increases. Combined with the combustion and reactivity test results of semi-coke, it can be seen that the reactivity of semi-coke is far higher than that of Lu'an anthracite, so the utilization efficiency of semi-coke in the blast furnace is higher, and the protection effect on coke is strengthened. After adding semi-coke for mixing and injection, the utilization efficiency of mixed pulverized coal in the furnace is improved, and the replacement ratio of coal to coke is increased. So, when the coal ratio increases after injecting semi-coke, the coke ratio will decrease and the fuel ratio will remain almost unchanged. Table 3-23 shows the changes in fuel costs (excluding taxes) during the baseline period and various stages of industrial testing.

Table 3-23 Changes in fuel cost of No. 3 blast furnace during the reference period and industrial trial period (yuan)

Item	Blended coal cost	Coke cost	Coal specific power consumption	Fuel cost
Base period	106.53	282.02	0	388.55
Stage 1	110.07	277.01	0.3	387.38
Stage 2	108.58	277.73	0.4	386.71
Stage 3	108.53	277.01	0.5	386.04

Note: In the calculation of fuel cost, Lu'an anthracite is calculated as 615.21 yuan/tHM on a dry basis (8% water is deducted), and semi-coke is calculated as 532.93 yuan/tHM on a dry basis (15% water is deducted). The mixed coal cost of fuel cost includes 179 gross taxes. The prices of coke and nut coke are calculated based on the internal settlement price of 680 yuan/tHM. According to the statistical results of data, when 10% semi-coke is added to the mixed coal, the power consumption cost of pulverizing powder by ton of iron coal injection increases by about 0.3 yuan, and when 20% is added, it is calculated as an increase of 0.4 yuan.

From Table 3-23, it can be seen that due to the increase in coal and the decrease in coke ratio after adding semi-coke injection, the cost of coal per ton of iron has increased, while the cost of coke has decreased. The grindability of semi-coke is lower than that of Lu'an anthracite, so the pulverizing cost will increase after adding semi-coke. Combining the change of blended coal cost per ton of iron, coke cost per ton of iron and pulverizing cost, the change of fuel cost in each stage of the reference period and test period can be calculated. The results show that compared to the reference period, when the proportion of semi-coke is 10%, the cost of iron fuel per ton decreases by 1.17 yuan. When adding 20% semi-coke, the fuel cost per ton of iron is reduced by 1.84 yuan. When adding 25% semi-coke, the fuel cost per ton of iron is reduced by 2.51 yuan. According to the current production of about 2,000t of pig iron per day in the No. 3 blast furnace, if the proportion of semi-coke is controlled at 20%, only the annual fuel cost of the No. 3 blast furnace can be reduced by about 1.3 million. If it can be controlled at 25%, the fuel cost can be reduced by about 1.8 million yuan. Moreover, due to the good stability of the raw fuel conditions, the blast furnace can maintain stable and smooth operation after injecting semi-coke. Therefore, the reduction in the cost of per ton of iron fuel will inevitably lead to a decrease in the cost of per ton of iron, this is also the purpose of this industrial experiment.

3.7 Analysis of energy saving and emission reduction effect of semi-coke injection

3.7.1 Energy consumption of ironmaking process

The ironmaking industry is an important basic industry of the national economy and has made great contributions to social and economic development. At the same time, the traditional ironmaking process is a process of intensive consumption of resources and energy. Its energy consumption accounts for more than 70% of the total energy consumption of steel products, and the pollution is very serious.

Iron and steel enterprises have always attached great importance to energy saving and consumption reduction. In 2010, the total energy consumption of China's key iron and steel enterprises was about 260 million tons of standard coal, up 6.189% from the previous year. The comprehensive energy consumption of tons of steel was 606.69kgce/t, among which, the energy consumption of blast furnace ironmaking process was about 407.58kgce/t, sintering process was about 55.27kgce/t, pelleting process was about 29.14kgce/t, coking process was about 109.10kgce/t, as shown in Fig. 3-31. Ironmaking system (including sintering, pellet, coking, blast furnace ironmaking) energy consumption accounts for about 70% of the total energy consumption of comprehensive iron and steel enterprises, the cost accounts for about 60%, pollutant emissions account for more than 70%, therefore, the iron and steel industry to reduce the comprehensive energy consumption of tons of steel must strive to reduce the energy consumption of ironmaking process. The premise of reducing the energy consumption of the ironmaking process is to reduce the ironmaking fuel ratio (coke ratio, small coke ratio and coal ratio). In 2010, the ironmaking fuel ratio of the national key iron and steel enterprises (excluding small coke ratio) was 518kg/tHM.

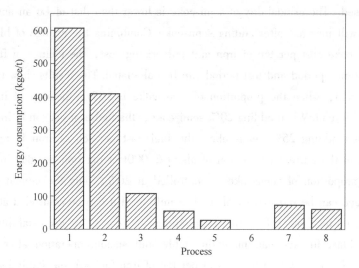

Fig. 3-31 Energy consumption of each process in key iron and steel enterprises in 2010
1—Ton steel comprehensive; 2—Blast furnace; 3—Coking; 4—Sintering;
5—Pelletizing; 6—Converter; 7—Electric furnace; 8—Steel rolling

More than 80% of the energy used by iron and steel enterprises is coal, mainly coking coal, fuel coal and blast furnace injection coal. The energy consumption of blast furnace ironmaking accounts for 49.4% of the energy consumption of iron and steel complex enterprises. 78% of the energy consumption of blast furnace comes from the combustion of carbon (coke and pulverized coal). The coal consumption of iron and steel industry accounts for more than 10% of the total coal consumption of the national industry. If the energy consumption of a ton of iron is 600kgce/t, each production of 100 million tons of iron needs to consume 0.6 million tons of standard coal. Therefore, in 2011, the energy consumption of China's iron making process alone was close to 400 million metric tons of coal, while the world's iron making process energy consumption exceeded 650 million metric tons of coal.

3.7.2 CO_2 emission of ironmaking process

Coal is composed of C, H, O and a small amount of N, S, P and other polymer organic compounds, in its combustion or consumption process will release a large number of CO_2, SO_2, NO_x, soot and other toxic and harmful substances, thus causing pollution to the environment. Among them, CO_2 is widely believed to be the most important contributor to the greenhouse effect and global warming greenhouse gas. In order to protect the earth's environment, the control of CO_2 emission has been a universal consensus in the world.

While the steel industry consumes a large amount of underground carbon sources, it also continuously emits CO_2 gas into the atmosphere. Its CO_2 emissions are second only to those of the power sector. If the CO_2 emission coefficient of standard coal is calculated as 2.3, the CO_2

emission of the molten iron production of 500 million tons molten iron is about 700 million tons, that is, every 100 million tons of molten iron produced by the iron industry will be discharged into the atmosphere about 140 million tons of CO_2. CO_2 emissions from the iron and steel industry account for about 7% of total anthropogenic CO_2 emissions. As a super steel country and a major greenhouse gas emitter, China is facing a severe emission reduction situation. In addition to emitting a large amount of CO_2, the ironmaking process also produces a large amount of harmful gases such as SO_2 and NO_x. In 2007, China's SO_2 emissions of ironmaking industry reached 7 million tons, NO_x emissions reached 3.22 million tons.

3.7.3 Effect of semi-coke powder injection on energy saving and emission reduction of blast furnace

3.7.3.1 Energy saving effect of semi-coke powder used in blast furnace injection

Semi-coke powder is mixed with blast furnace injection coal and injected into blast furnace, and complex physical and chemical reaction is carried out in the tuyere and raceway. Then the ash enters the slag and participates in the slagging process and desulfurization reaction. The thermal effect in the process of semi-coke powder injection is mainly reflected in the comprehensive effect of combustion heat release and ash heat consumption, desulfurization heat consumption and decomposition heat consumption. In addition to providing direct heat, the indirect effects of combustibility and reactivity of semi-coke powder on blast furnace can not be ignored. The combustibility and reactivity of semi-coke determine the output of unburned coal powder, which will affect the material column permeability, unfavorable to furnace condition, mainly in the following aspects: (1) the unburned coal powder enters the slag in a suspended state, which will increase the viscosity of the slag, affect the fluidity of the slag, and cause the accumulation of the hearth in severe cases. (2) unburned coal powder will stay in the soft melting zone and dripping zone, reducing its permeability, resulting in the lower part of the difficult and suspended material. (3) a large amount of unburned coal powder is absorbed on the surface of the charge and deposited in the void, especially in the central part will seriously deteriorate the permeability of the charge, resulting in an increase in pressure differential, central airflow blocking, edge airflow development and bad furnace conditions. The fluctuation of blast furnace condition will inevitably lead to load reduction in operation, fuel ratio will rise, and energy consumption of blast furnace will increase.

The semi-coke powder injection into blast furnace undergoes a series of processes such as volatile decomposition (hydrocarbon decomposition), water gas reaction and volatile, combustion, fixed carbon combustion, desulfurization, and ash change into slay. These processes are accompanied by many endothermic reactions. In the blast furnace, the final products after combustion of semi-coke powder are CO, N_2 and H_2; Therefore, the actual heat supply of semi-coke in the blast furnace should be the heat after deducting the heat of decomposition heat, water gas reaction heat, desulfurization heat consumption and slag producing heat of semi-coke from the heating value of incomplete combustion of carbon in semi-coke. Generally, the calorific value measurement of semi-coke includes the heat release from hydrogen combustion, but in fact, hydrogen from the semi-coke is not oxidized in the blast furnace, that is to say, high hydrogen content does not increase the heat contribution to the combustion in the tuyere. Therefore, in the actual heat supply calculation of semi-coke, the final product of the reaction involving hydrogen is hydrogen, which does not burn and release heat. The specific calculation process is shown in Fig. 3-32.

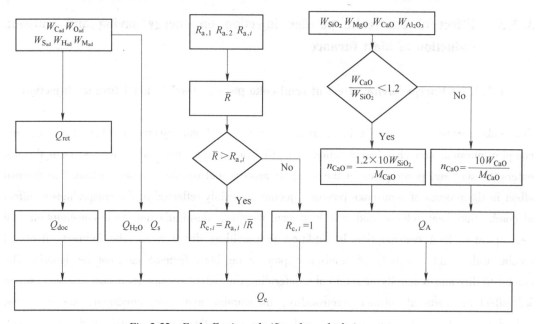

Fig. 3-32 Fuel effective calorific value calculation process

In this section, the concept of effective heating value proposed by experimental research is used to compare and analyze the effect law of injecting semi-coke on the heat of blast furnace. The effects of semi-coke on energy saving are analyzed from decomposition heat consuption, slag heat consumption, desulfurization heat consumption and so on.

This comparative analysis uses Hengyuan semi-coke powder and Yangquan anthracite for comparative analysis. The proximate analysis and ultimate analysis of the two fuels are shown in Table 3-24.

Table 3-24 Fuel composition analysis

Sample	Proximate analysis				Ultimate analysis				
	A_{ad}	V_{ad}	FC_{ad}	M_{ad}	C_{ad}	H_{ad}	N_{ad}	$S_{t,ad}$	O_{ad}
Hengyuan semi-coke powder	10.19	9.88	79.29	0.64	80.87	2.35	0.82	0.34	4.79
Yangquan anthracite	9.95	7.06	82.36	0.63	83.29	3.31	1.12	0.62	6.66

By theoretical calculation, the specific values of desulfurization heat consumption, slag-

3.7 Analysis of energy saving and emission reduction effect of semi-coke injection

forming heat consumption, decomposition heat consumption and water gas reaction heat consumption of Hengyuan semi-coke and Yangquan anthracite injected into blast furnace can be obtained. According to the calculation process in Fig. 3-32, the effective heating value of different fuels can be calculated.

Since the sulfur content of Hengyuanlan semi-coke powder is 0.34%, which is only half of Yangquan anthracite (0.62%), replacing Yangquan anthracite with Hengyuanlan semi-coke powder can reduce the sulfur load in furnace and reduce the heat consumption of desulfurization. As can be seen from Table 3-25, the heat consumption of desulfurization of Hengyuan semi-coke powder is 28.56kJ/kg, about half that of Yangquan anthracite. From the point of view of desulphurization heat consumption, Hengyuan semi-coke powder saves 45.16% heat consumption compared with Yangquan anthracite.

Table 3-25 Fuel effective heating value and each heat   (kJ/kg)

Coal type	Desulfurization heat consumption	Slag forming heat consumption	Decomposition heat consumption	Heat consumption of water gas reaction	Effective heating value
Hengyuan semi-coke powder	28.56	360.12	261.54	44.27	7,591.31
Yangquan anthracite	52.08	351.64	680.43	43.58	6,845.99
Consumption reduction (%)	+45.16	-2.41	+61.56	-1.58	+10.89

After low temperature heat treatment, some volatiles of Hengyuan semi-coke powder are removed, and the hydrogen content in carbon matrix is reduced, which is lower than that of Yangquan anthracite. Through theoretical calculation, the decomposition heat consumption of Hengyuan semi-coke powder is 261.54kJ/kg, and the decomposition heat consumption of Yangquan anthracite is 680.43kJ/kg. From the perspective of decomposition heat consumption, Hengyuan semi-coke powder will save 61.56% heat than Yangquan Yuan anthracite. The heat expenditure of different fuels is shown in Fig. 3-33.

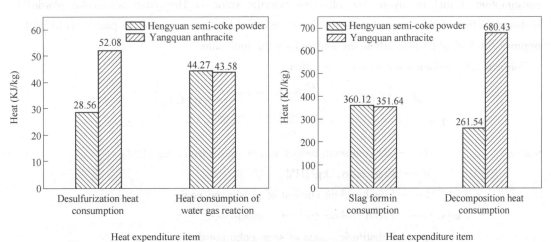

Fig. 3-33 Comparison of head expenditure of different fuels

However, the ash content and moisture content of Hengyuan semi-coke powder are slightly higher than those of Yangquan anthracite, resulting in heat consumption in slag formation and water gas reaction consumption. The slag-forming heat consumption and water-gas reaction heat consumption of Hengyuanlan semi-coke powder are 2.41% and 1.58% higher than that of Yangquan anthracite, respectively. However, the difference of heat consumption of these two items is much smaller than that of desulphurization heat consumption and decomposition heat consumption, which has little influence on the final effective calorific value.

It can be seen from the combustion characteristics of Table 3-26 that the combustion characteristics of Hengyuan semi-coke is slightly better than that of Yangquan anthracite, and the initial ignition temperature of Hengyuan semi-coke is significantly lower than that of Yangquan anthracite. Combined with the combustion properties of the two fuels and the income and expenditure of different heat, the effective calorific value of Hengyuan semi-coke is 7,591.31kJ/kg, and that of Yangquan anthracite is 6,845.99kJ/kg. The effective calorific value of Hengyuan semi-coke is 745.32kJ/kg higher than that of Yangquan anthracite. From the perspective of effective heat of fuel, the effective heat value of Hengyuan semi-coke powder is 10.89% higher than that of Yangquan anthracite. Hengyuan semi-coke powder has the possibility of replacing Yangquan anthracite for injection.

Table 3-26 Analysis of calorific value and combustibility of fuel

Coal type	Combustion rate (%)			lower calorific value (kJ/kg)
	500℃	600℃	700℃	
Hengyuan semi-coke powder	17.73	50.36	79.93	29,218
Yangquan anthracite	7.12	43.41	79.90	31,065

3.7.3.2 Emission reduction effect of semi-coke powder in blast furnace injection

Semi-coke powder is used instead of anthracite for blast furnace injection, which can reduce the consumption of fuel in tuyere. The effective calorific value of Hengyuan semi-coke powder is higher than that of Yangquan anthracite. The injection of Hengyuan semi-coke powder is helpful to improve the fuel utilization efficiency and reduce the fuel ratio.

Ton iron CO_2 emission reduction calculation:

$$Y = \frac{M \cdot \gamma \left(w(C)_C - \dfrac{Q_{\text{Effective heat of coal}}}{Q_{\text{Effective heat of semi-coke}}} \cdot w(C)_S \right)}{m_C} \times m_{CO_2}$$

where
- Y——CO_2 emission reduction per ton of iron, kg/tHM;
- M——Coal ratio, kg/tHM;
- $w(C)_C$——Fixed carbon content of pulverized coal;
- $w(C)_S$——Fixed carbon content of semi-coke;
- γ——Substitution ratio of semi-coke powder;
- $Q_{\text{Effective heat of coal}}$——Effective calorific value of injected coal, MJ/Kg;

$Q_{\text{Effective heat of semi-coke}}$ —— Effective calorific value of semi-coke, MJ/Kg;

m_C —— Relative atomic mass of carbon, 12;

m_{CO_2} —— Relative molecular mass of carbon dioxide, 44.

Ton iron SO_2 emission reduction calculation:

$$Y = \frac{M \cdot \gamma \left(w(S)_C - \dfrac{Q_{\text{Effective heat of coal}}}{Q_{\text{Effective heat of semi-coke}}} \cdot w(S)_S \right)}{m_S} \times m_{SO_2} \times \lambda$$

where
- Y —— SO_2 emission reduction per ton of iron, kg/tHM;
- M —— Coal ratio, kg/tHM;
- $w(S)_C$ —— Sulfur content of pulverized coal;
- $w(S)_S$ —— Sulfur content of semi-coke;
- γ —— Substitution ratio of semi-coke powder;
- $Q_{\text{Effective heat of coal}}$ —— Effective calorific value of injected coal, MJ/Kg;
- $Q_{\text{Effective heat of semi-coke}}$ —— Effective calorific value of semi-coke, MJ/Kg;
- m_C —— Relative atomic mass of sulfur, 32;
- m_{CO_2} —— Relative molecular mass of sulfur dioxide, 64;
- λ —— Sulfur dioxide emission ratio (Accounting for the sulfur load in the furnace).

It can be obtained from the above calculation that the effective calorific value of Hengyuan semi-coke powder is 7,591.31kJ/kg, that of Yangquan anthracite is 6,845.99kJ/kg, and that of sulfur discharged with blast furnace top gas accounts for about 5% of sulfur load into the blast furnace. In this calculation process, semi-coke is used to replace 25% of blast furnace coal injection, and the enterprise coal injection into is 150kg/tHM. The specific calculation process is as follows:

Ton iron CO_2 emission reduction calculation:

$$\frac{150 \times 0.8329 - 150 \times \dfrac{6,845.99}{7,591.31} \times 0.8087}{12} \times 0.25 \times 44 = 14.24 \text{kg/tHM}$$

Ton iron SO_2 emission reduction calculation:

$$\frac{150 \times 0.0062 - 150 \times \dfrac{6,845.99}{7,591.31} \times 0.0034}{32} \times 0.25 \times 64 \times 0.05 = 0.012 \text{kg/tHM}$$

The calculation results show that the replacement of 25% Yangquan anthracite with Hengyuan semi-coke powder for injection will result in a emission reduction of 14.24kg of CO_2 and 0.012kg of SO_2 per ton of iron.

3.7.3.3 Energy saving advantage of semi-coke powder used in blast furnace injection

In view of the energy demand of ironmaking process and the need of CO_2 emission reduction, according to the above research, it can be judged that using semi-coke resources to assist ironmaking is a good choice. Although semi-coke energy can not completely replace all kinds of energy in ironmaking process, especially coke in the blast furnace to play the role of stock column skeleton. However, compared with traditional anthracite injection, semi-coke assisted ironmaking has great advantages.

A Good flammability and reactivity

The results show that the combustion and gasification reactivity of semi-coke are better than that of anthracite. If the semi-coke powder is used for blast furnace injection, the semi-coke powder can react rapidly in the raceway. This can reduce the amount of unburned carbon or increase the amount of injection. If large semi-coke and traditional coke are added directly into the blast furnace, the semi-coke can be consumed in the upper part of the blast furnace without affecting the gas permeability of the blast furnace. If the semi-coke reaches the lower part of the blast furnace, it can replace the solution loss reaction of some coke, thus reducing the coke ratio. In addition, the semi-coke powder can be quickly vaporized in the blast furnace to absorb heat, so as to reduce the temperature of the heat reserve area of the blast furnace, improve the utilization efficiency of gas, and realize the low reducing agent operation of the blast furnace.

B Clean and environmentally friendly

The semi-coke is cleaner than the traditional coal, and its harmful element content is lower. It can be used to produce high-quality direct reduced iron or high-quality clean steel materials. The sulfur content of semi-coke powder is very low, which can reduce the sulfur load of blast furnace, and then reduce the heat consumption of desulfurization in blast furnace, which is helpful to reduce the fuel consumption, iron making cost and CO_2 emission. In addition, the semi-coke powder resources are relatively clean, the content of N and S is lower than that of pulverized coal, which can greatly reduce the emission of pollutants such as SO_2 in the traditional iron making process, and realize the environmental friendliness of the iron making process to a greater extent.

C Comprehensive and efficient use of resources, sustainable development

The utilization direction of semi-coke powder in semi-coke products is the least, and most of the semi-coke power are stacked in the production plants. Therefore, the rational utilization of semi-coke powder has become one of the key problems that the semi-coke industry needs to solve. Semi-coke powder is used to assist ironmaking, which can be used for blast furnace injection. The waste semi-coke powder resources can be used as a heating agent and reducing agent, which can realize the full utilization of resources and reduce environmental pollution. The application of semi-coke powder in blast furnace injection is a breakthrough connection between two different industries, which can realize the optimal allocation and efficient utilization of resources, help reduce the dependence of blast furnace injection on high-quality bituminous coal, maximize the utilization of the byproducts of semi-coke industry, and truly reflect the strategic goal of sustainable development.

3.8 Chapter summary

(1) Semi-coke is a carbocoal product in the coal chemical industry, with high fixed carbon,

high specific resistance, high chemical activity, low ash, low aluminum, low sulfur, low phosphorus characteristics. Compared with ordinary anthracite, semi-coke has better combustion performance, but its grindability index is relatively low, which has a certain influence on the pulverizing ability of the mill. In order to ensure good grindability of semi-coke in the process of injection, raw coal with low cohesiveness and low pyrolysis temperature are generally selected for production. Increasing the granularity of the injection fuel into the furnace is helpful to relieve the pulverizing pressure of the mill, but it has certain influence on the combustion of the fuel the tuyere. Iron and steel enterprises need to match the grindability of semi-coke and the combustion rate in the tuyere.

(2) Semi-coke can be used to replace high quality anthracite resources for blast furnace injection. The commonly used collocation structure is mixed injection of semi-coke and bituminous coal. From the perspective of composition, the high volatile content and high moisture content of bituminous coal contribute to the formation of carbonaceous pore structure and promote combustion. From the perspective of micromorphology, the lamellar and grooved micromorphology of bituminous coal has a better effect on combustion promotion than that of stone-shaped semi-coke. From the perspective of chemical structure, the content of fatty side chain and oxygen-containing functional group of bituminous coal is higher than that of semi-coke, which is also an important factor leading to better combustion performance of bituminous coal than semi-coke. Therefore, the mixed injection of bituminous coal and semi-coke can further improve the combustion efficiency.

(3) The sulfur content and nitrogen content of semi-coke powder are very low. Its use can reduce the sulfur load of blast furnace, and then reduce the heat consumption of desulfurization in blast furnace. It is helpful to reduce the burnup, reduce the cost of iron making and reduce the CO_2 emission. Using semi-coke powder in blast furnace injection can realize the optimal allocation and efficient utilization of resources between two different industries, help to reduce the dependence of blast furnace injection on high-quality anthracite and maximize the utilization of semi-coke products, and truly reflect the strategic goal of sustainable development.

(4) The technology of injecting semi-coke powder into blast furnace of Xinxing Ductile Iron Pipes Co., LTD. was successful, and the cost per ton of iron was significantly reduced. At present, the technological achievements have been promoted and applied in well-known domestic enterprises such as Baotou Iron and Steel Co., LTD., Jiujiang Iron and Steel Co., LTD., Shanxi Meijin Iron and Steel Co., LTD., bringing 147 million yuan of economic benefits to iron and steel enterprises.

References

[1] Lin Jinyuan. Application of blue charcoal in calcium carbide production [J]. Chemical Technology and Economics, 2004, 22 (12): 23-25.

[2] Jiao Yang, Hu Binsheng, Gui Yong liang. Feasibility analysis of coal injection for blast furnace in JISCO [J]. Metallurgical Energy, 2011, 30 (6): 20-22.

[3] He Xinjie, Zhang Jianliang, Qi Chenglin, et al. Effect of catalyst on combustion characteristics and kinetics of pulverized coal [J]. Iron and Steel, 2012 (7): 74-79.

[4] Huo W, Zhou Z J, Chen X L, et al. Study on CO_2 gasification reactivity and physical characteristic of biomass, petroleum coke and coal chars [J]. Bioresour Technol, 2014, 159: 143-149.

[5] Lahijani P, Zzinal Z A, Mohamed A R, et al. CO_2 gasication reactivity of biomass char: catalytic influence of alkaline earth and transition metal salts [J]. Bioresour Technol, 2013, 144: 285-288.

Chapter 4　Basic Research on Blast Furnace Biomass Injection Technology

4.1　Overview of biomass

Biomass can refer to all living substances on earth. According to the view of energy resources, biomass is usually defined as the general term of a certain amount of accumulated animal and plant resources and wastes from animals and plants (excluding fossil resources). Therefore, biomass includes many kinds, agricultural and forestry resources such as crops, wood, animal manure, waste paper, sewage treatment plant residual sludge and many industrial wastes are also considered biomass.

Biomass is one of the most promising renewable energy sources, with various types and huge reserves, which is a green and environmentally friendly carbon-containing energy. Due to its potential availability, it can solve the world energy crisis to a certain extent[1-2]. In addition, the use of biomass can also effectively reduce global warming and pollution problems, compared with fossil energy, biomass has outstanding advantages in terms of economy and environmental protection. Therefore, the rational utilization of biomass has become a hot topic in recent years.

At present, domestic research on biomass ironmaking mainly focuses on the combustion weight loss characteristics of biomass, simple preparation of biomass char and energy consumption environmental assessment, etc. There are relatively few systematic studies on the application of agricultural and forestry wastes to blast furnace injection, especially few studies on the behavior of unburned residual carbon in the blast furnace generated in the tuyere raceway after biomass char is used for blast furnace injection[3-4]. In addition, most of the foreign research results are based on charcoal powder injection, which is not suitable for the current severe environmental situation in China [5-10]. As a potential injection fuel for ironmaking, the development of reasonable utilization technology of agricultural and forestry waste in blast furnace injection process is of great practical significance to improve the efficiency of high-temperature combustion of biomass/pulverized coal, reduce the dependence of ironmaking system on fossil fuels, and broaden the ideas of energy saving and reduction of CO_2 emissions in iron and steel enterprises.

Biomass used as fuel for blast furnace injection must meet the physical, chemical and basic properties of the incoming fuel. From a physical point of view, biomass must be crushed and screened to a particle size suitable for injection, and from a chemical point of view, biomass must have similar chemical components and combustion reaction properties as coal for injection. However, the composition and structure of biomass are very different from pulverized coal. Compared to pulverized coal, biomass has a lower calorific value, fixed carbon content, grindability and energy density, a larger volume, and a higher moisture and volatile content, all of which largely limit the direct and effective application of biomass in blast furnace injection processes. In the last decade, pyrolysis has been developed as a promising thermochemical technology for the reforming of biomass under inert gas or anoxic conditions. Compared with raw biomass, the solid product obtained from biomass pyrolysis is called biomass char, which has lower moisture content, higher carbon content, higher calorific value and energy density, and can meet the requirements of fuel for blast furnace injection under certain pyrolysis conditions.

4.2 Study on physicochemical properties of biomass char

4.2.1 Experimental materials and preparation

The experiment used three agricultural wastes common in China, namely corn cob, palm shell and soybean straw. Among them, the corn cob and soybean straw are taken from the northeast region of our country, and the palm shell is taken from the Guangxi Zhuang autonomous region in south of our country. The Northeast China has vast plains with deep, fertile black soil and is the largest commercial grain production base in China, of which corn and soybeans are the main cash crops. As a by-product, corn cob and soybean straw have huge resources and great potential use value. In addition, the Northeast China has the highest level of agricultural mechanization in the country, which provides convenient conditions for the recycling of biomass resources. Palm shell is as a common agricultural waste in the south, which itself has a higher calorific value than general biomass and has a higher economic value.

Before the biomass pyrolysis, the three biomass feedstocks were first dried in a drying oven at 318K for 2h, and then crushed into pellets using a crusher, and the pellet diameter was controlled to be below 1mm.

The pyrolysis equipment of biomass char is a medium temperature tube resistance furnace, the structure of which is shown in Fig. 4-1. The tube furnace is programmed and the heating process is divided into two steps: first from room temperature to 573K with a heating rate of 5K/min; then from 573K to the target temperature with a heating rate of 10K/min and a holding time of 3h. The pyrolysis temperatures for each group were 1,073K, 1,273K and 1,473K. Each time, 50g of sample was weighed into the graphite field and placed horizontally at the bottom of the furnace tube with tongs, and then heated in a tube furnace. N_2 was used as a protective gas during the pyrolysis to ensure that the biomass samples were pyrolyzed under air-insulated conditions, and the gas flow rate was set at 2L/min. The pyrolysis time was 30min,

Fig. 4-1 Medium temperature tube furnace equipment

after which the furnace tube was removed and cooled naturally to room temperature, and N_2 was continuously introduced during the cooling process to ensure the accuracy of the experiment. Finally, the cooled biomass char was taken out and put into the specimen bag.

The char prepared in the tube furnace was further crushed and the biomass char fines below 200 mesh (75μm) were screened for experiments and testing. For the convenience of exposition, corn cob char, palm husk char and soybean straw char are indicated by CC char, PS char, and SS char in the graphs, respectively.

4.2.2 Biomass char yield

Fig. 4-2 shows the yields of the three biomass char at different pyrolysis temperatures, from which it can be seen that the three groups of samples have the same trend, i.e., the yield gradually decreases with the increase of pyrolysis temperature. When the pyrolysis temperature increased from 1,073K to 1,273K and 1,473K, the yield of corn cob char was 23.40%, 20.83%, 20.42%, soybean straw char yield (mass fraction) was 25.95%, 24.06%, 21.79% and palm shell char yield (mass fraction) was 33.12%, 29.22%, 28.18%. At the same pyrolysis temperature, the three biomass char yields were, in descending order, palm shell char > soybean straw char > corn cob char. In addition, it can be seen from Fig. 4-2 that the yield of palm shell char and corn cob char at 1,273K is significantly lower than that at 1,073K. When the pyrolysis temperature increases to 1,473K, the yield changes very little compared to 1,273K. This indicates that the pyrolysis process of palm shell char and corn cob char is basically completed at 1,273K. The yield of soybean straw char showed a good linear relationship with the pyrolysis temperature, and the yield of char decreased gradually with the increase of the pyrolysis temperature, indicating that the pyrolysis process was still going on in the interval of 1,073-1,473K.

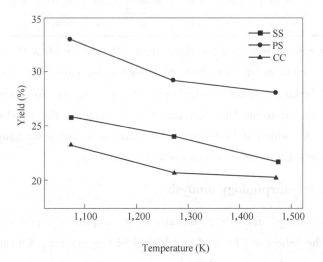

Fig. 4-2 The relationship between yield and pyrolysis temperature

4.2.3 Proximate and ultimate analysis

Like pulverized coal composition, biomass char is composed of four components: moisture, volatile matter, ash and fixed carbon. Based on the proximate and ultimate analysis, the composition characteristics of different biomass char can be initially determined and thus their industrial use can be determined.

Table 4-1 shows the results of proximate and ultimate analyses for all samples at dry basis, where d represents the dry basis. The fixed carbon content of all biomass char samples reached above 70%, and the ash content gradually increased and volatile content gradually decreased with the increase of pyrolysis temperature. Palm shell char had the highest ash content and the lowest corresponding fixed carbon content at the same pyrolysis temperature, while soybean straw char had the highest volatile content among the three biomass chars, which may be related to the hemicellulose and cellulose content of the biomass feedstock.

Table 4-1 Proximate and ultimate analysis of different biomass char (wt. %)

Sample	Pyrolysis temperature (K)	Proximate analysis (%)			Ultimate analysis (%)					H/C atomic ratio
		FC_d	A_d	V_d	C	H	O	N	S	
CC char	1,073	86.13	5.25	8.62	84.42	1.565	13.130	0.70	0.185	0.0185
	1,273	87.62	5.58	6.80	84.89	1.175	13.022	0.79	0.123	0.0138
	1,473	89.08	5.67	5.25	86.15	0.681	12.174	0.86	0.135	0.0079
PS char	1,073	76.22	10.72	13.06	79.76	2.227	17.102	0.82	0.091	0.0279
	1,273	72.76	20.05	7.19	75.80	1.122	22.154	0.83	0.076	0.0148
	1,473	72.52	23.33	4.15	74.18	0.618	24.286	0.84	0.076	0.0083
SS char	1,073	72.11	9.04	18.85	72.73	2.627	22.475	1.39	0.778	0.0361
	1,273	78.76	11.59	9.65	76.98	1.198	20.381	1.13	0.311	0.0156
	1,473	80.73	12.30	6.97	75.96	0.900	21.611	1.19	0.339	0.0118

From Fig. 4-3, it can be seen that the three groups of samples show the same trend, with a gradual decrease in H/C as the pyrolysis tempreature increases. The main reason for this phenomenon is that the large amount of volatile fraction precipitation takes away most of the H and O elements in the raw material. The decrease in H/C implies the development of aromatic hydrocarbon structures, which indicates that the increase in pyrolysis temperature favors the development of aromatic hydrocarbon structures in biomass char.

4.2.4 Microscopic morphology analysis

The microscopic morphology analysis of biomass char samples was performed with a Hitachi SU8010 SEM from the School of Chemical and Biological Engineering, University of Science and Technology Beijing, which has excellent low acceleration voltage imaging capability, 1kV secondary electron resolution up to 1.3nm, and is equipped with high stereo image and high

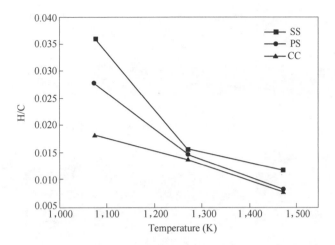

Fig. 4-3 H/C curve of biomass char

resolution SE and BSE image signal detectors. In this experiment, the microscopic morphology characteristics of corn cob char, palm shell char and soybean straw char prepared at three different pyrolysis temperatures (1,073K, 1,273K, 1,473K) were observed. The SEM images of the three sets of biomass char samples were shown in Fig. 4-4. Due to the small particle size of the biomass char, the magnification of the shot was set to 2,000×in order to better observe the microscopic pore structure changes of the biomass char.

The type and structure of the biomass feedstock have a great influence on the microscopic morphology of high temperature pyrolysis biomass char as can be seen in Fig. 4-4. The microscopic shapes of the biomass char were generally consistent with the raw materials. Among them, the particles of palm shell char were irregular blocks under microscopic observation. When the pyrolysis temperature was 1,073K, the surface of palm shell char was distributed with dotted micropores as well as long shallow trace-like pores, and the surface was smooth without melting loss. This indicates that at 1,073K, the pore structure of the palm shell char is still in the development stage. When the pyrolysis temperature increases to 1,273K, the development of pores and the fusion phenomenon between pores can be clearly observed from the SEM images, and the strip-like pores are spread all over the surface of the palm shell char, and the size of the pores increases significantly compared with the sample at 1,073K. This is because with the increase of pyrolysis temperature, the further precipitation of volatile fraction makes the pore structure of palm shell char develop, the pore wall becomes thinner, the pore volume increases, and due to the continuous development of pores, the fusion of adjacent pores will occur at the relatively weak pore wall. When the pyrolysis temperature further increases to 1,473K, the surface melting of palm shell char is obvious, the pores collapse and blockage, and the pore structure disappears. When the pyrolysis temperature increased from 1,073K to 1,273K, the pore structure of palm shell char increased. When the pyrolysis temperature increased from 1,273K to 1,473K, the pore structure of palm shell char decreased sharply, and the overall pore structure

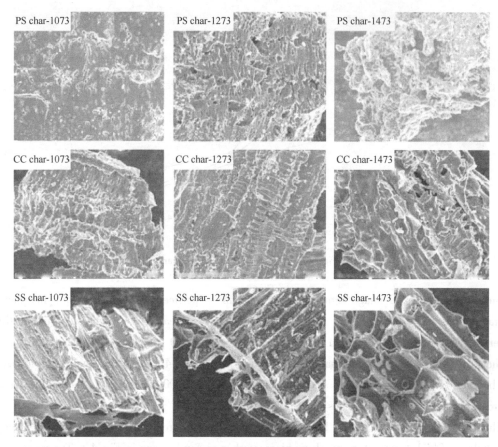

Fig. 4-4 SEM images of different biomass chars

showed a trend of increasing and decreasing first.

Corn cob char particles have a laminar skeleton structure, and at 1,073K, the surface of corn cob char is distributed with circular open pits and rough surface with prismatic protrusions of different lengths, which is less pore structure compared with palm shell char. When the temperature increased to 1,273K, the surface of corn cob char was melted, most of the spherical open pits disappeared, the prismatic protrusions were shortened and part of the skeleton structure disappeared, and the surface of the particles was smoother than at 1,073K due to the melting phenomenon. When the temperature increased to 1,473K, the melting and erosion phenomenon further intensified, and the surface pits and prismatic skeleton of corn cob char almost completely disappeared, and the surface smoothness further increased. The pore structure of the corn cob char was influenced by the pyrolysis temperature, and its own pore structure gradually decreased with the increase of the pyrolysis temperature.

The microscopic shapes of soybean straw char differed significantly from those of palm shell char and corn cob char, showing a hollow tubular structure of thin sheets. This is because soybean straw is composed of rhizomes that serve as water and nutrient transporters. The soybean char prepared by pyrolysis at 1,073K has almost no pore structure other than hollow tubular pores, accompanied by striated protrusions on the surface; The microscopic morphology of soybean char prepared by pyrolysis at 1,273K was less affected by the pyrolysis temperature, except for the slight development of tubular pores, which did not change significantly. When the pyrolysis temperature increases to 1,473K, the holes melt, the elliptical holes develop into prismatic holes, the wall thickness decreases sharply, and circular penetration holes appear between the walls of adjacent holes, which indicates that the fusion between adjacent holes begins to occur. The pore structure of soybean straw char generally showed an increasing trend with pyrolysis temperature.

4.2.5 Specific surface area and pore structure analysis

The microscopic pore structure is one of the important factors affecting the combustion and gasification reactivity of biomass char. Previous studies have shown that pore structures with diameters larger than 1.5nm (1.7-200nm) promote the char gas curing reaction. In order to further determine the specific values of specific surface area and pores and to quantitatively analyze the differences in microstructure of biomass char, the experiments were performed using an Ouadrasorb SI type specific surface area analyzer from Conta, USA, with an inert gas N_2 as the adsorbent. The sample was degassed at 473K for 6h, and then the adsorption and desorption amounts were measured at fixed specific pressure points for the specimen. 77K adsorption specific pressure points were taken from 0.01 to 0.995 for 7 points, while 19 points were taken during the desorption process. Under the conditions of nitrogen adsorption, the specific surface area of the adsorbent was calculated by the BET (Brunauer-Emmett-Teller) method, and the average pore size and pore volume were calculated by the BJH (Barret t-Johner-Halenda) method.

Adsorption and desorption isotherms are the basic source of data for the study of adsorption and desorption phenomena: surface and pore structure of materials. The type of material itself, the surface topography, the pressure and temperature of the adsorbent all influence the process of adsorption and desorption. A large amount of information about the pore structure can be obtained from the adsorption and desorption isotherms. In 1985, IUPAC published a manual on *Report on Physical Adsorption Data for Gas/Solid Systems* with special reference to the determination of surface area and porosity, and its conclusions and recommendations have been widely accepted by science and industry and are used today. The isothermal adsorption-desorption curves of different biomass char in this experiment were shown in Fig. 4-5, where the horizontal coordinate is the ratio of the test pressure to the N_2 saturation pressure, i.e., the relative pressure point p/p_0, and the vertical coordinate is the pore volume. As can be seen in Fig. 4-5, the adsorption and desorption curves of all samples are of type II curves, which are characterized by an overall inverse S-shape. In the low relative pressure stage, the curves rise slowly, which implies that the adsorption process goes from monolayer to multilayer and also indicates that more micropores are present in the samples. In the high relative pressure stage, the curves rise rapidly, showing the presence of medium and macroporous structures. In addition, the adsorption and desorption curves were not completely closed at the lower phase of the low relative pressure. The main reason is the presence there of a large number of very narrow slit pores or bottle-shaped pores, where the N_2 molecules move very slowly at 77K, so the adsorption of very narrow pores is kinetically limited. Such curves indicate that the pore system in all biomass char in this experiment is continuous and intact, and the pore size range is widely distributed, from molecular level to uncapped open pores simultaneously.

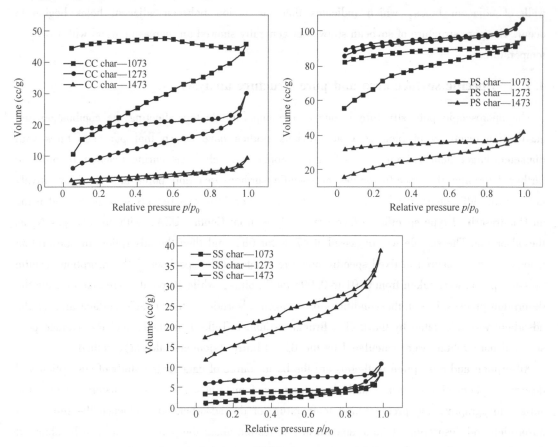

Fig. 4-5　Isothermal adsorption and desorption curves of different biomass char

The specific surface areas of different biomass char were shown in Table 4-2. The specific surface areas were very different, among which the palm shell char had the largest specific surface area with 237.36m^2/g, 408.18m^2/g and 75.45m^2/g at 1,073K, 1,273K and 1,473K, respectively, while the corn cob char had 71.66m^2/g, 35.97m^2/g and 7.48m^2/g, 35.97m^2/g and 7.48m^2/g, and the specific surface of soybean straw char was 6.19m^2/g, 6.58m^2/g, and 54.83m^2/g. The evolution pattern of the specific surface area of different biomass char was different, among which the specific surface area of palm shell char increased and then decreased with the increase of pyrolysis temperature, which is consistent with the SEM images observed that the number of pores of palm shell char increased with the increase of pyrolysis temperature and the expansion of intrinsic pore volume led to the increase of specific surface area. With the further increase in temperature, the pores undergo fusion collapse, and the fusion of the particle surface and pore plugging lead to a sharp decrease in their own specific surface area. The specific surface area of corn cob char gradually decreased with the increase of pyrolysis temperature, which was due to the gradual disappearance of the inherent pore structure on the surface with the increase of pyrolysis temperature, and the pore fusion phenomenon gradually increased with the increase of pyrolysis temperature. The specific surface area of soybean straw char showed a gradual increase with the increase of pyrolysis temperature in general. When the pyrolysis temperature was not higher than 1,273K, the microscopic morphology of soybean straw char was less affected by the change of pyrolysis temperature, and its own pore structure only slightly increased, so when the pyrolysis temperature increased from 1,073K to 1,273K, the specific surface area only increased by 0.39m^2/g. When the pyrolysis temperature increased to 1,473K, the specific surface area of soybean straw char suddenly increased dramatically, which was due to the rapid development and expansion of the pores of the particles at 1,473K.

Table 4-2 Specific surface area of different biomass char   (m^2/g)

Pyrolysis temperature (K)	SS char	CC char	PS char
1,073	6.19	71.66	237.36
1,273	6.58	35.97	408.18
1,473	54.83	7.48	75.45

The pore size distribution of different biomass char samples were shown in Fig. 4-6. The pores can be classified into three categories according to the pore size: micropores, mesopores and macropores. The pore size of micropores is less than 2nm, the pore size of mesopores is 2-50nm, and the pore size of macropores is more than 50nm. It can be seen that the pore size distribution within the biomass char particles is more uniform and the majority of the pore structure is mesoporous. With the increase of pyrolysis temperature, the pore size distribution of the particles changed significantly, and the changes mainly occurred in the mesoporous part. The pore size distribution of palm shell char decreased and then increased with the increase of pyrolysis temperature, while the pore size distribution of corn cob char showed a gradual decrease, and the pore size distribution of soybean straw char increased slightly with the increase of pyrolysis temperature and then suddenly increased at 1,473K. The variation in this part is mainly due to the pore development, fusion and collapse caused by volatile analysis out and carbon fusion erosion. The beginning of the curve is an ascending part, and as the pore size increases, a spike appears in the 3.5-4nm region, indicating that this pore size range dominates the mesopore structure. The pore size distribution and specific surface area of these three groups of biomass char have the same variation pattern. When the specific surface area increases, the number and proportion of mesopores increases and vice versa, and it can be seen that the specific surface area of mesopores with pore size less than 40nm accounts for most of the total specific surface area of biomass char.

4.2.6 Microcrystalline structure analysis

Although biomass is mainly composed of cellulose, hemicellulose and lignin, the nature and

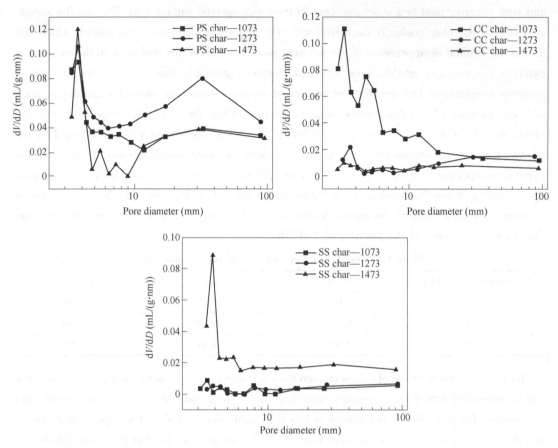

Fig. 4-6 Pore size distribution curves of different samples

content of the three components vary greatly among different types of organisms. Under high-temperature pyrolysis, the O—H, C—H and C—O bonds within and between molecules are broken in large numbers, and aromatic structures are formed at the same time, resulting in changes in the arrangement and stacking of the microstructure of pyrolyzed char. Since Warren applied X-ray diffraction (XRD) to coal samples, XRD has been widely used for the detection and analysis of graphite-like crystalline material in carbonaceous materials such as coal coke. The microcrystalline structure of biomass char consists of several layers of aromatic lamellae stacked on top of each other, and the length of the lamellae is often referred to as the lamellae diameter, denoted by L_a. The distance between the aromatic lamellae is denoted by d_{002}; The average stacking thickness of each microcrystal lamellae (i.e., the microcrystal height) is denoted by L_c. The specific values of these microcrystalline parameters can be obtained using calculations such as the Bragg equation and the Scherrer formula, which in turn clearly characterize the differential variation in the microcrystalline structure of born-matter samples.

The test equipment is a RigakuD MAX2500PC X-ray diffractometer from the State Key Laboratory of New Metallic Materials, University of Science and Technology Beijing, and the test conditions are: Cu radiation, X-ray tube voltage of 35kV, X-ray tube current of 40mA; The scanning angle is 10°-100°, the scanning speed is 2°/min, the sampling interval is 0.02°, and the sample size is less than 74um. The results were shown in Fig. 4-7.

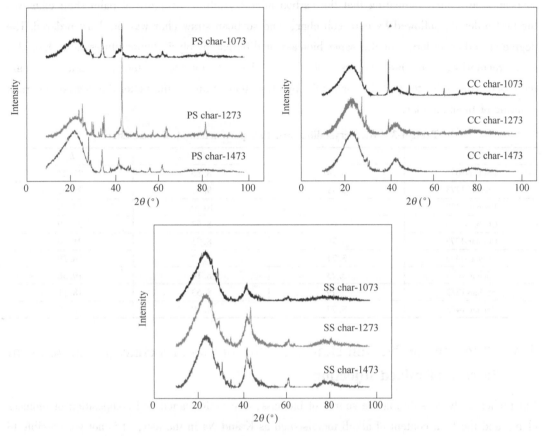

Fig. 4-7 XRD spectra of different samples

From Fig. 4-7, it can be observed that other crystalline characteristic peaks appear near the (002) and (100) peaks of each group of samples, which is due to the relatively high composition of biomass char ash prepared after high-temperature pyrolysis, and inorganic minerals will produce characteristic peaks in the X-ray diffraction process without de-ashing treatment.

The peak intensity and diffraction angle of the characteristic peak (002 peak) were taken into the equation of Scherrer et al. to calculate the unit structure size of the carbon microcrystal:

$$L_c = 0.89\lambda/(\beta_{002}\cos\theta_{002}) \qquad (4-1)$$

$$d_{002} = \lambda/(2\sin\theta_{002}) \qquad (4-2)$$

$$L_a = 0.89\lambda/(\beta_{100}\cos\theta_{100}) \qquad (4-3)$$

where, β is the half-height width of the (002) peak; the X-ray wavelength λ is 0.15405nm; θ is the diffraction angle of the 002 peak; L_c, L_a and d are the stack height, microchip layer length, and crystal layer spacing of the microcrystalline lamellae, respectively.

The stacking height of the microcrystalline lamellae reflects the degree of orientation of the aromatic layers in spatial arrangement, while the length of the lamellae reflects the degree of condensation of the aromatic rings. From Table 4-3, it can be seen that the L_c values of different biomass char differed at the same pyrolysis temperature, i.e., palm shell char > corn cob char > soybean straw char, indicating that the carbon microcrystalline structure of palm shell char was the best ordered, followed by corn cob char, and soybean straw char was the least ordered. The degree of carbon ordering of the same biomass at different pyrolysis temperatures also showed a clear regularity, with the increase of the pyrolysis temperature, the L_c value gradually increased. This indicates that the higher the pyrolysis temperature, the better the degree of carbon ordering of biomass char.

Table 4-3　Microcrystalline structure parameters of biomass char

Sample	d_{002}	L_c	L_a
PSchar-1073	3.74	8.86	23.21
PSchar-1273	3.71	9.61	23.62
PSchar-1473	3.69	10.48	24.59
CCchar-1073	3.67	8.67	22.29
CCchar-1273	3.70	8.95	21.20
CCchar-1473	3.71	9.02	20.39
SSchar-1073	3.73	8.14	19.56
SSchar-1273	3.79	8.53	18.11
SSchar-1473	3.77	8.70	16.20

4.3　Study on combustion characteristics of biomass char/pulverized coal in case of mixed injection

Comprehensively considering the source of biomass, the performance and composition of biomass char, and the high content of alkali metals such as K and Na in the ash, it is not yet possible to completely use biomass char for blast furnace injection, so it can be considered to mix biomass char with injection coal to replace part of anthracite for injection. Biomass char is mainly used in the combustion reaction in front of the tuyere after blast furnace injection, so the mixed combustion characteristics of biomass char and pulverized coal are particularly important.

At present, the methods for studying blast furnace injection biomass or biomass char mainly include thermogravimetric analyzer, fluidized bed, dropper furnace, etc., among these technologies, thermogravimetric analysis method has been widely used because of its simple experimental process and rich experimental data. The characteristic parameters of combustion reaction include starting combustion temperature, burnout temperature, maximum combustion rate and corresponding temperature, average combustion rate and combustion time, etc., which can be obtained by thermogravimetric analysis. At the same time, kinetic research is also crucial for the design of mixed combustion reactions, and reaction kinetic parameters such as activation energy, pre-exponential factor and reaction mechanism functions can also be obtained by analyzing the conversion rate corresponding to different temperatures. And the combustion information obtained by thermogravimetric analysis can predict and evaluate the reaction behavior of fuels in large-scale equipment.

4.3 Study on combustion characteristics of biomass char/pulverized coal in case of mixed injection

Compared with corn cob and soybean straw, palm shell has the characteristics of rich yield, high yield and excellent performance, so the char prepared from palm shell and Yangquan anthracite coal were selected for mixed combustion experiment, the mixed combustion reaction process and characteristic parameters were analyzed, and the influence of the dosing amount of palm shell char on the mixed combustion process and combustion characteristic parameters was studied. The kinetic model of combustion reaction was established, the kinetic parameters of mixed combustion reaction were solved, the results of different kinetic models were compared and analyzed, and the synergistic law of mixed combustion of palm shell char and pulverized injection coal was explored, which provided a theoretical basis for the reasonable combination of palm shell char and pulverized coal for blast furnace injection.

4.3.1 Experimental materials and methods

Before the experiment, the coal sample was placed in a dry drying box at 105℃, and after 12h, it was taken out and ground in a sealed sample mill, and the pulverized coal particles with a particle size of less than 0.074mm were taken out by a standard sieve for backup. In this experiment, palm shell char were prepared by pyrolysis at 600℃.

Table 4-4 is the chemical composition of pulverized coal, palm shell original and palm shell char, it can be seen from the table that the volatile content of palm shell is obviously high dry pulverized coal, but the fixed carbon content is much lower than that of pulverized coal, only 26.78%. After pyrolysis, the volatile content of palm shell decreases sharply and the fixed carbon content increases. From the ultimate analysis, it can be seen that the H and O element content of palm shell decreases after pyrolysis, while C increases significantly. The molar ratio of H/C and O/C of palm shell is greater than that of pulverized coal, mainly due to the relatively low aromatic content of palm shell. After pyrolysis, the volatile content, ash and fixed carbon content of pulverized coal and palm shell char are not much different, but the content of harmful elements N and S of palm shell char is lower than that of pulverized coal.

Table 4-4 Chemical composition of palm shell, palm shell char and pulverized coal

Sample	Industrial Analytics (%)				Ultimate analysis (%)					Molar ratio	
	M_{ad}	V_{ad}	A_{ad}	FC_{ad}	C_{ad}	H_{ad}	O_{ad}	N_{ad}	S_{ad}	H/C	O/C
Palm shell	4.92	67.71	1.91	25.46	44.62	5.15	42.74	0.47	0.19	1.39	0.72
Palm shell char	1.84	7.51	10.83	79.82	80.82	1.92	3.68	0.82	0.08	0.29	0.03
Pulverized coal	0.84	7.90	12.32	78.94	79.86	2.94	2.26	1.11	0.67	0.44	0.03

In order to investigate the effects of palm shell char dosage and heating rate on the combustion characteristics of palm shell char and pulverized coal, the prepared biomass char and pulverized coal were mixed according to the ratio of 20%, 40%, 60% and 80% of the mass fraction of palm shell char, and ground in an agate mortar for 10min until it was fully mixed, and then packed into bags for subsequent thermal analysis experiments, and the heating rate during the experiment was set to 5℃/min, 10℃/min, 15℃/min and 20℃/min, respectively. The specific experimental protocol is shown in Table 4-5.

Table 4-5 Experimental protocol of mixed combustion of pulverized coal and palm shell char

Experiment serial number	Experiment serial number	Heating rate(℃/min)
1	Pulverized coal	5/10/15/20
2	20% palm shell char+ 80% pulverized coal	5/10/15/20
3	40% palm shell char+ 60% pulverized coal	5/10/15/20
4	60% palm shell char+ 40% pulverized coal	5/10/15/20
5	80% palm shell char+ 20% pulverized coal	5/10/15/20
6	Palm shell char	5/10/15/20

In this experiment, a thermogravimetric analyzer was used to determine the combustion characteristics of palm shell char and pulverized coal and their mixtures. The sample weighed (5.0±0.1) mg was evenly added to the corundum crucible (ϕ5mm×5mm), placed on the differential thermal balance, the program was set to heat up, so that the sample was heated to 800℃ under an air atmosphere, the equipment automatically recorded the weightlessness process, and collected data, and the TG-DTG curve of the combustion process of the experimental sample was drawn by Origin drawing software after the experiment.

In order to explore the reasons for the difference in combustion characteristics between palm shell char and pulverized coal, the analysis method introduced in Section 4.2 was used to study the physical and chemical characteristics of palm shell char and pulverized coal, and the difference between the physical and chemical properties of the two was analyzed, so as to explore the mechanism of mixed combustion of palm shell char and pulverized coal.

4.3.2 Mixed combustion characteristics and synergistic effect of biomass char/pulverized coal

The thermogravimetry of combustion reaction of palm shell char, pulverized coal and its mixture under different conditions was studied by gravimetric analysis, and the combustion characteristics were analyzed, and the corresponding combustion characteristic parameters were described.

4.3.2.1 Biomass char and pulverized coal combustion characteristics

The combustion characteristics of a single substance were analyzed at a heating rate of 20℃/min, and Fig. 4-8 showed the TG-DTG curves of palm shell char and pulverized coal, respectively.

Comparing Fig. 4-8, it can be found that the combustion process of palm shell char and pulverized coal is roughly the same, both have only one main combustion reaction stage, and some key combustion characteristic parameters (such as starting combustion temperature T_i, burnout temperature T_f, maximum combustion rate and peak R of DTG curve and their corresponding peak temperature T_m) are used to compare the combustion characteristics of palm shell half-sale and pulverized coal, and the results are listed in Table 4-6. As can be seen from the table, the starting combustion temperature of palm shell char is 388℃, which is significantly lower than that of pulverized coal (501℃), indicating that palm shell char is more likely to catch fire than pulverized coal. The burnout temperatures of palm shell char and pulverized coal were 534℃ and 649℃, respectively, indicating that the combustion process of palm shell char ended earlier than pulverized coal, mainly because the alkali metal content in the char ash of palm shell was high, which played a catalytic role in the later stage of the combustion reaction. The DTG curves of palm shell char and pulverized coal have only one obvious weight loss peak, the difference is reflected in the sharp and narrow weight loss peak of the coal dust DTG curve, and the peak temperature is 599℃. The weight loss rate peak of palm shell char is wide, located between 490-520℃. The study by Mundike et al.[11] found that the DTG curve of mimosa char prepared at 578℃ had a similar phenomenon. The maximum combustion rate of pulverized coal is $311×10^{-3} s^{-1}$, and the maximum combustion rate of palm shell char is $2.55×10^{-3} s^{-1}$.

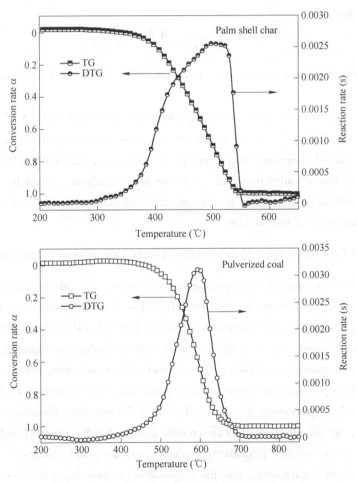

Fig. 4-8 Variation curve of conversion rate and reaction rate of palm shell char and pulverized coal combustion with temperature

Table 4-6 Characteristic parameters of mixed combustion of palm shell char and pulverized coal

Palm shell char dosage (%)	T_i (℃)	T_{m1} (℃)	$R_1 \times 10^{-3}$ (s)	T_{m2} (℃)	$R_2 \times 10^{-3}$ (s)	T_f (℃)	$R_{0.5} \times 10^{-4}$ (s)
0	501	—	—	599	3.11	649	2.85
20	458	—	—	586	2.77	638	2.94
40	417	507	1.49	585	2.18	632	3.09
60	402	502	1.85	566	1.57	605	3.28
80	391	501	2.31	—	—	566	3.41
100	388	497	2.55	—	—	534	3.52

Note: R_1 and R_2 are the first and second peaks on the DTG curve, respectively; T_{m1} and T_{m2} are the peak corresponding temperatures.

According to the combustion TG-DTG curve of palm shell char and pulverized coal in Fig. 4-8, in order to evaluate the combustion reactivity of palm shell char and pulverized coal, the reactivity characteristic parameter $R_{0.5}$ is used to characterize the combustion reactivity of the experimental sample:

$$R_{0.5} = \frac{0.5}{t_{0.5}} \quad (4-4)$$

where, $t_{0.5}$ is the time required for the conversion rate of the combustion reaction to reach 0.5, s.

The reactivity characteristic parameters $R_{0.5}$ of palm shell chars are $3.52 \times 10^{-4} \, s^{-1}$, the pulverized coal is $2.85 \times 10^{-4} \, s^{-1}$, and the reactivity characteristic parameters of palm shell char are greater than that of pulverized coal, indicating that the combustion reactivity of palm shell char is better than that of pulverized coal, from the above analysis, it can be seen that mixing palm shell char with pulverized coal can promote the combustion of pulverized coal and improve the combustion reactivity of the mixture.

4.3.2.2 Effect of biomass char dosage on mixed combustion characteristics

The percent of biomass char is an important factor affecting the mixed combustion process of pulverized coal and biomass char, and the TG-DTG curve of mixed combustion of palm shell char and pulverized coal with different dosages is shown in Fig. 4-9 under the condition of 20℃/min heating rate. It can be seen from the figure that the mixed combustion characteristics depend on the individual combustion characteristics of each fuel, the combustion curve of the mixture is between the separate combustion curve of palm shell char and pulverized coal, and with the gradual increase of the proportion of palm shell char in the mixture, the shape of each curve is slightly different, specifically related to the content added, but the overall trend is that with the palm shell char in the mixture proportions gradually increase, the combustion curve gradually moves to the low temperature area, and the combustion characteristics of the mixture and palm shell char are getting closer and closer. It is clear from the DTG curve that the peak of DTG lies between the separate combustion intervals of pulverized coal and palm shell char. The increase in

4.3 Study on combustion characteristics of biomass char/pulverized coal in case of mixed injection

the amount of palm shell char in the mixed fuel causes the combustion curve to shift to the left, indicating that the temperature at which the combustion reaction occurs gradually decreases.

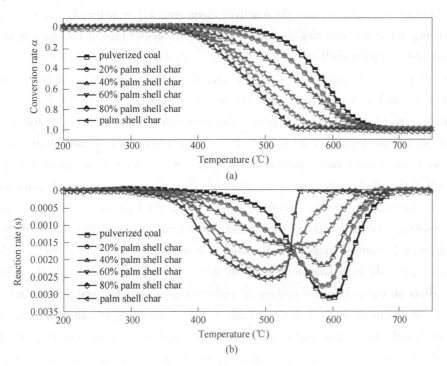

Fig. 4-9 Mixed combustion curve of palm shell char and pulverized coal
(a) conversion curve; (b) reaction rate curve

When the proportions of palm shell char are 40% and 60%, the DTG curve of mixture combustion appears two peaks, the first peak is due to the combustion of palm shell char, located between 200-550℃, and the second peak is due to the combustion of pulverized coal, located between 550-700℃. When the addition amount of palm shell char in the mixed fuel is small (e.g. 20%), its combustion characteristics are closer to pulverized coal, and the combustion weight loss peak belonging to palm shell char is difficult to distinguish. Similarly, when the dosing amount of palm shell char is 80%, the combustion behavior of the mixed fuel is closer to that of palm shell char, and the combustion peak attributed to pulverized coal is difficult to identify. When the percent of palm shell char is higher than 50%, the peak of the DTG curve is located in the low temperature region, and the peak value gradually increases with the increase of the palm shell char addition to the low temperature region. When the dosing amount of palm shell char is less than 50%, the peak of the DTG curve is located in the high temperature region, and the peak value gradually decreases with the increase of palm shell char dosage to the low temperature region. Mundike et al.[11] found the same pattern when they studied the mixed combustion characteristics of different biomass char and pulverized coal. In order to accurately study the influence of the dosing amount of palm shell char on the combustion process of the mixture, the characteristic parameters of the combustion process of pulverized coal and palm shell char and the mixture were calculated, which are listed in Table 4-6.

Fig. 4-10 and Fig. 4-11 show the relationship curve between the starting combustion temperature T_i, the burnout temperature T_f and the reactivity characteristic parameters $R_{0.5}$ and the palm shell char percent, it can be seen that the initial combustion temperature of the mixture is between pulverized coal and palm shell char, and decreases with the increase of palm shell char dosage, for the mixture with 80% palm shell char addition, the initial combustion temperature is reduced by 110℃ compared with pulverized coal. The reason for the decrease in the initial combustion temperature is mainly that the starting combustion temperature of the palm shell char is low, and the palm shell char begins to burn at a lower temperature, and releases heat, so that the temperature of the mixture rises, promoting the release and combustion of volatiles in pulverized coal, so that the combustion process is advanced [12]. With the increase of palm shell char dosage, the burnout temperature decreased significantly, the burnout temperature of pulverized coal was 649℃, while the burnout temperature of the mixture with 80% palm shell char reason for the decrease in burnout temperature is mainly due to the presence of alkali metal oxides in the chars ash of palm shells. A large number of studies [13-14] show that alkali metal oxides have a catalytic effect on the combustion process of carbonaceous materials, and the catalytic effect is enhanced with the increase of alkali metal oxide content, and the K_2O content in palm shell char ash is significantly higher than that of pulverized coal, and with the increase of palm shell char dosage, the catalytic effect of K_2O is enhanced, resulting in a gradual decrease in the burnout temperature of the mixture. With the gradual increase of the char percent of palm shell, the reactivity characteristic parameter $R_{0.5}$ gradually increased, indicating that the addition of palm shell char improved the combustion reactivity of pulverized coal.

It can be seen from the mixed combustion curve of palm shell char and pulverized coal that the combustion process becomes more complicated after mixing palm shell char and pulverized coal, and with the increase of the amount of palm shell char in the mixture, the combustion

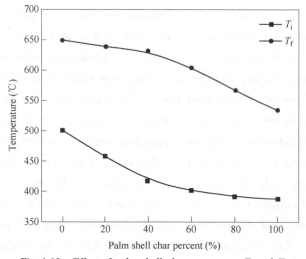

Fig. 4-10 Effect of palm shell char percent on T_i and T_f

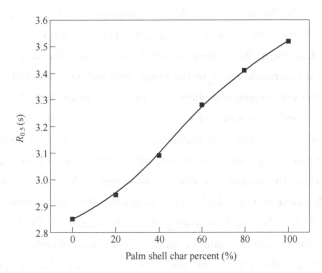

Fig. 4-11 Relationship between palm shell char percent and reactivity characteristic parameters

characteristic parameters of the mixed fuel (such as the starting combustion temperature T_i, burnout temperature T_f and peak temperature T_m) do not show linear addition, which indicates that there may be a synergistic effect during the mixed combustion of palm shell char and pulverized coal. In order to explore whether there is a synergistic effect between palm shell char and pulverized coal in the mixed combustion process, the theoretical conversion rate of the mixed combustion process of palm shell char and pulverized coal was calculated and compared with the experimental data. If the experimental data is higher than the theoretical data, it indicates that there is a positive synergy between the half-sale of palm shell and the mixed combustion of pulverized coal; when the experimental data is lower than the theoretical data, there is a negative synergy between them[15]. When the experimental data and theoretical data are the same, it indicates that there is no synergy in the mixed combustion process of the two. The synergy in the mixed combustion process is directly related to the difference in experimental data and theoretical data of mixed combustion of palm shell char and pulverized coal.

Assuming that the palm shell char and pulverized coal do not interact during the mixed combustion process, and the conversion rate of the mixture is the weighted average of the conversion rate of the two components burned separately, it can be calculated by Eq. (4-5):

$$\alpha_{cal} = a \cdot \alpha_{PC} + b \cdot \alpha_{YQ} \tag{4-5}$$

where, a and b are the mass fractions of palm shell char and pulverized coal in the mixture, respectively; α_{PC} and α_{YQ} are the conversion rates of palm shell char and pulverized coal burned separately, respectively.

The comparison of experimental data and theoretical data of the mixed combustion process is shown in Fig. 4-12. As can be seen from the figure, the experimental data and theoretical data are close, but there are still differences in different combustion stages. In the low temperature stage, after the reaction starts, when the conversion rate is relatively low, the experimental data are significantly lower than the theoretical data. As the conversion rate increases, the two sets of data get closer and closer. In the high temperature stage, the experimental data are significantly higher than the theoretical data when the conversion rate is large. From the above analysis, it can be inferred that there is a synergistic effect in the mixed combustion process of palm shell char and pulverized coal, in the low temperature stage, the mixing of palm shell char and pulverized coal limits the volatilization and combustion of volatile components, anthracite coal ranks are high, and higher temperatures are required to make the volatile analysis and ignite: when the palm shell char and pulverized coal particles are mixed, the energy transmission is affected by particles of different sizes, so that the temperature gradient increases, which causes the volatilization and combustion of volatiles to move to the high temperature area. With the increase of the proportion of palm shell char plus ratio, the negative synergy gradually weakened. The high temperature stage is mainly the combustion of carbon residues, and the combustion of carbon in the palm shell char releases heat and heats the carbon in the pulverized coal, thereby promoting its early combustion. At the same time, the alkaline substances in the chars ash of palm shells also act as a catalyst to promote the combustion of pulverized coal, and the positive synergy gradually increases with the increase of the proportion of palm shell char compounding, which may be because the content of alkaline substances in the mixture increases with the increase of the proportion of palm shell char compounding.

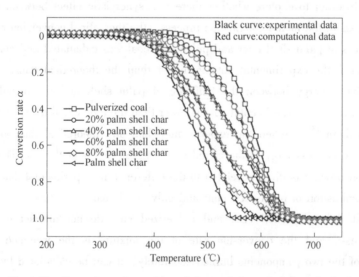

Fig. 4-12 Comparison of experimental data and theoretical data of mixed combustion of palm shell char and pulverized coal

Farrow et al. [16] found that the removal of alkali metals and alkaline earth metal compounds in biomass will lead to the almost complete disappearance of the synergistic effect in the mixed combustion process of biomass chars and pulverized coal, which further proves the catalytic effect of alkali metals and alkali earth metals contained in biomass on pulverized coal combustion. The study by Edreis et al. [17] shows that alkali metals and alkaline earth metals in biomass ash are key factors in the interaction between biomass and pulverized coal mixture gasification during gasification. Due to the positive effect of biomass char particles, the combustion characteristics of pulverized coal are better than the corresponding weighted average performance parameters, resulting in the synergistic effect of mixed combustion of biomass char and pulverized coal. Wang et al. [18] also found similar results to this experiment when studying the co-gasification reaction of petroleum coke and biomass char mixtures.

Therefore, from the above analysis, it can be concluded that in the mixed combustion process of palm shell char and pulverized coal, the presence of pulverized coal inhibits the volatilization and combustion of the volatile content of palm shell char, but the combustion of palm shell char and the alkali metal and alkaline earth metal in the char ash obviously promote the ignition and combustion of pulverized coal. Literature [19] also reached similar conclusions when studying the mixed combustion of papermaking waste residue and municipal solid waste.

4.3.2.3 Effect of heating rate on mixed combustion characteristics

The TG-DTG curves of the mixed combustion experiment of palm shell char and pulverized coal under different heating rates are shown in Fig. 4-13, and it can be seen that the hybrid combustion characteristic curves of palm shell char and pulverized coal with different proportions have the same trend with the heating rate. It is roughly shown that with the increase of the heating rate, the TG and DTG curves both move to the high temperature area, and the more obvious the combustion peaks on the DTG curve, the corresponding peak temperature also shows an increasing trend. The weight loss peak of the DTG curve is higher, that is, the maximum combustion rate increases, indicating that the heating rate increases the combustion intensity. This phenomenon is consistent with the findings of other researchers[20]. This is because during the combustion process, the greater the heating rate, the stronger the thermal shock of the mixed

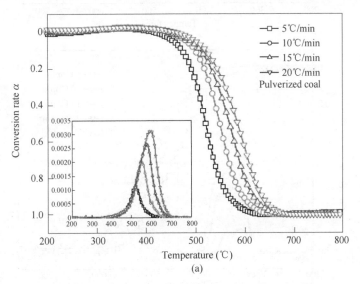

(a)

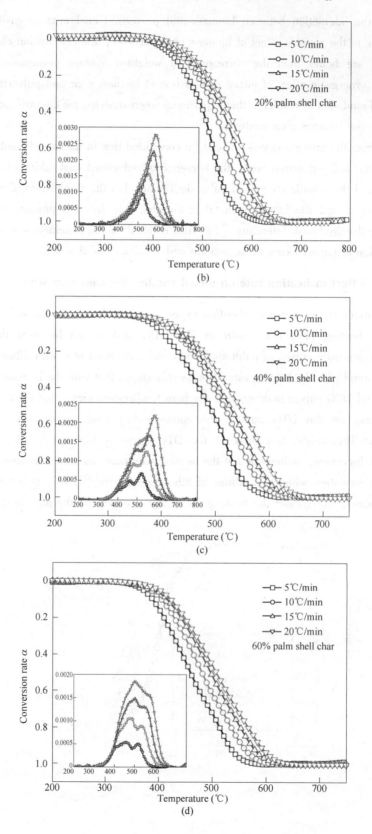

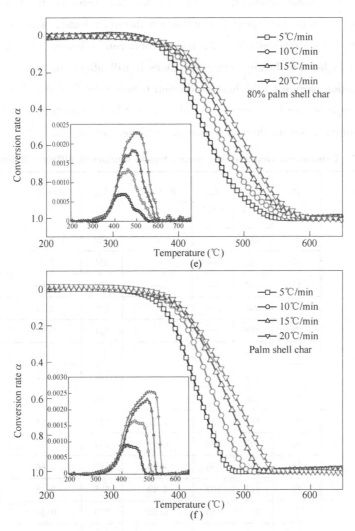

Fig. 4-13 Mixed combustion characteristics curve of palm shell char and pulverized coal at different heating rates

sample in a short period of time, which accelerates the reaction rate of the sample with oxygen, and the concentrated precipitation of moisture, volatile matter and fixed carbon in a short period of time.

The characteristic parameters of combustion reaction under different heating rate conditions are shown in Table 4-7, the change of starting combustion temperature T_i, burnout temperature T_f and $R_{0.5}$ with the heating rate is shown in Fig. 4-14 and Fig. 4-15, with the increase of heating rate, the starting combustion temperature T_i and burnout temperature T_f gradually increase, and the combustion reactivity characteristic parameter $R_{0.5}$ gradually increases, indicating that with the increase of heating rate, the combustion reactivity of the mixture is enhanced. There are two main reasons for this phenomenon: on the one hand, the higher the heating rate, the faster the sample burns to the same temperature, the shorter the reaction residence time at this temperature, and the lower the degree of reaction. On the other hand, when the heating rate is low, the fuel particles are gradually heated, and the heat can be better transferred to and around the particles, the higher the heating rate, the more it will affect the heat conduction in and around the combustion particles, the heat conduction between the device and the sample, so the temperature rises rapidly, and the burning sample does not have enough time to reach the thermal equilibrium, resulting in severe thermal hysteresis at higher heating rates [21-22].

Table 4-7 Combustion characteristic parameters of samples at different heating rates

Palm shell char percent (%)	β (℃/min)	T_i (℃)	T_{m1} (℃)	$R_1 \times 10^{-3}$ (s)	T_{m2} (℃)	$R_2 \times 10^{-3}$ (s)	T_f (℃)	$R_{0.5} \times 10^{-4}$ (s)
0	5	456	—	—	522	1.13	604	0.80
0	10	482	—	—	551	2.00	629	1.52
0	15	490	—	—	579	2.66	649	2.20
0	20	501	—	—	599	3.11	665	2.85
20	5	419	—	—	521	0.95	598	0.81
20	10	435	—	—	546	1.59	625	1.55
20	15	439	—	—	578	2.26	644	2.25
20	20	458	—	—	586	2.77	655	2.94
40	5	382	461	0.51	521	0.68	589	0.85
40	10	402	478	0.93	543	1.24	613	1.62
40	15	408	496	1.30	557	1.66	630	2.38
40	20	417	507	1.49	585	2.18	649	3.09
60	5	371	454	0.55	520	0.52	561	0.91
60	10	385	487	1.08	537	1.89	589	1.73
60	15	393	501	1.49	548	1.30	606	2.51
60	20	402	502	1.85	566	1.57	622	3.28
80	5	370	450	0.70	—	—	533	0.95
80	10	377	463	1.34	—	—	556	1.82
80	15	380	494	1.82	—	—	571	2.63
80	20	391	501	2.31	—	—	583	3.41
100	5	359	419	0.93	—	—	476	0.99
100	10	375	451	1.66	—	—	500	1.87
100	15	378	493	2.32	—	—	518	2.71
100	20	388	497	2.55	—	—	542	3.52

4.3 Study on combustion characteristics of biomass char/pulverized coal in case of mixed injection

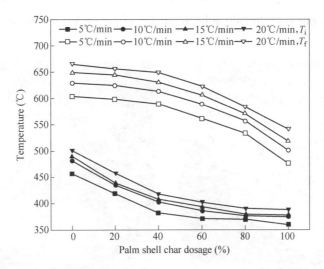

Fig. 4-14 The relationship between different heating rates, starting combustion temperature T_i, burnout temperature T_f and palm shell half-joule dosage

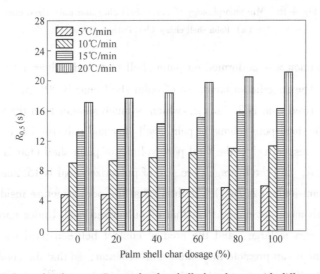

Fig. 4-15 Relationship between $R_{0.5}$ and palm shell char dosage with different heating rate reactivity characteristic parameters

4.3.2.4 Physical and chemical properties of biomass char and pulverized coal

There are great differences in the chemical composition and microstructure of palm shell char and pulverized coal, and studying their physical and chemical properties will help to analyze and compare the differences in combustion characteristics. Fig. 4-16 shows the microscopic morphology of palm shell char and pulverized coal observed by scanning electron microscopy, both of which are irregular spherical particles, the surface of pulverized coal is relatively smooth, the structure is compact, and there are almost no holes. However, there are a large number of holes on the surface of palm shell char, mainly due to the large number of volatiles escaping during the pyrolysis preparation process, forming a large number of pore structures on the surface and inside of the particles.

Fig. 4-16 Micromorphology of palm shell char and pulverized coal
(a) Palm shell char; (b) Pulverized coal

N_2 adsorption detection was performed on palm shell char and pulverized coal, the results are shown in Table 4-8. The specific surface area of palm shell char is 201.5m²/g, and the specific surface area of pulverized coal is 3.3m²/g, which is much smaller than the specific surface area of palm shell char; the total pore volume of palm shell char and pulverized coal is 136.8×10^{-3} cm³/g and 126×10^{-3} cm³/g, respectively, the total pore volume of palm shell char is more than 10 times that of pulverized coal, and the average pore size of pulverized coal is 15.2nm, which is 7 times that of palm shell semi-joule (2.7nm). It shows that the pore structure inside the char particles of palm shell is much more developed than that of pulverized coal. Under normal circumstances, the specific surface area is large, and the area of contact between particles and air during the reaction is large, which can promote the combustion reaction, so that the combustion reactivity of palm shell char is significantly greater than that of pulverized coal [23].

Table 4-8 Palm shell char and pulverized coal N_2 adsorption results

Sample	Specific surface area (m²/g)	Total pore volume×10^{-3} (cm³/g)	Average aperture (nm)
Palm shell char	201.5	136.8	2.7
Pulverized coal	3.3	12.6	15.2

X-ray diffraction and Raman spectroscopy were used to analyze the difference of carbon structure of palm shell char and pulverized coal, Fig. 4-17 shows the X-ray diffraction spectrum of palm shell char and pulverized coal, it can be seen that the diffraction angle corresponding to the (002) peak of pulverized coal is large dry palm shell char, and the (002) peak of pulverized coal is narrower and sharper than the peak of palm shell semi-joule. Table 4-9 shows the calculated microcrystalline structural parameters of palm shell char and pulverized coal, the microcrystalline stacking height L_c and the number n of microcrystalline stacking layers of palm shell char are smaller than that of pulverized coal, while the microcrystalline layer spacing d_{002} is greater than that of pulverized coal, indicating that the degree of carbon ordering and graphitization of palm shell char is smaller than that of pulverized coal, resulting in the combustion reactivity of palm shell char being better than that of pulverized coal.

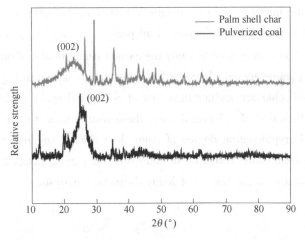

Fig. 4-17 XRD pattern of palm shell char and pulverized coal

Table 4-9 Microcrystalline structural parameters of palm shell char and pulverized coal

Sample	Angle (°)	Half height wide (°)	L_c (nm)	d_{002}(nm)	n
Palm shell char	23.41	7.52	1.066	0.379	3.80
Pulverized coal	25.45	3.60	2.236	0.350	7.39

Raman spectroscopy is a powerful detection technique because it is sensitive not only to crystal structure, but also to molecular structure, and is widely used to characterize almost all carbonaceous materials[24-25]. Raman spectra of carbonaceous materials are usually divided into primary and secondary regions, and for perfect graphite, there is only one peak (called G peak) that appears at 1,580cm^{-1} in the primary region, which corresponds to a tensile vibration mode with E_{2g} symmetry in the aromatic layer of the graphite crystal. For highly disordered carbon, additional peaks caused by defects in the micro-lattice appear in the primary regions of 1,150cm^{-1}, 1,350cm^{-1}, 1,530cm^{-1} and 1,620cm^{-1}[26]. The peak at 1,350cm^{-1} (D_1) is often referred to as the defect peak, corresponding to the graphite lattice vibration mode with A_{1g} symmetry and attributed to in-plane defects such as defects and heteroatoms [27]. When peak D_1 is present, a peak located at 1,620cm^{-1} (peak D_2) always appears, and its intensity decreases with increasing order[28]. The peak (D peak) located at 1,530cm^{-1} is usually a very wide band around 1,500-1,550cm^{-1}, which is considered to be derived from carbon produced by amorphous sp^2 bond hybridization, such as organic molecules, fragments or functional groups, which exist in poorly organized material structures [29], which may be related to reaction sites and carbon reactivity. The peak (D_4 peak) located at 1,150cm^{-1} appears in very poorly ordered materials such as soot and coal coke [30], and its attribution is still debated.

Fig. 4-18 shows the Raman spectrum of palm shell char and pulverized coal, and it can be seen that the intensity of G peak and D peak of pulverized coal is large and dry palm shell char. The Raman spectral peaks of palm shell char and pulverized coal were divided into 1 Lorentz peak (G) and 4 Gaussian peaks (D_1, D_2, D_3, D_4) by Peakfit peak fitting software, and the fitting accuracy reached more than 0.99, and the fitting results are shown in Fig. 4-19 of half-height width (FWHM-G), the intensity ratio of troughs V and G peaks between D peaks and G peaks (I_V/I_G), the ratio of G peak area to the sum of all peak areas (A_G/A_{all}) and the order of carbon structure [30-31], which is widely used to study the carbon structure of carbonaceous materials. The Raman characteristic parameters calculated by fitting results are shown in Fig. 4-20, the FWHM-G and I_V/I_G of palm shell char are greater than that of pulverized coal, while the A_G/A_{all} of palm shell char is smaller than that of pulverized coal, these results indicate that the degree of carbon structure ordering and graphitization degree of palm shell char is smaller than that of pulverized coal, resulting in the combustion reactivity of palm shell char being better than that of pulverized coal, which is consistent with the results of X-ray diffraction analysis.

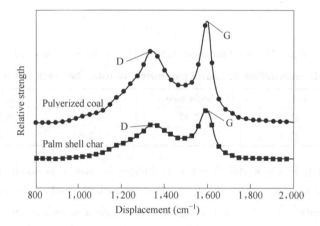

Fig. 4-18 Raman spectra of palm shell char and pulverized coal

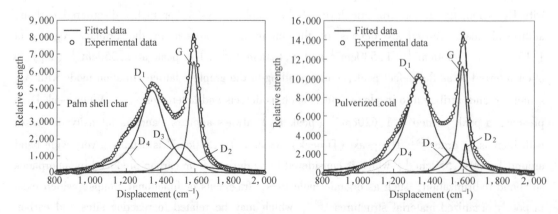

Fig. 4-19 Comparison of peak fitting curve and experimental curve of palm shell char and pulverized coal Raman pattern

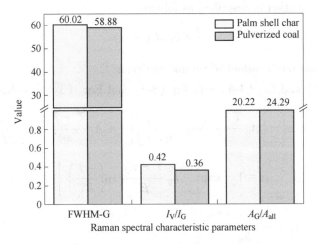

Fig. 4-20 Comparison of characteristic parameters of
palm shell char and pulverized coal Raman pattern

4.3.3 Kinetic analysis of biomass char/pulverized coal mixed combustion

4.3.3.1 Kinetic model

In this section, two commonly used gas-solid reaction kinetic models are used to study the combustion kinetic process of palm shell semi-coke, pulverized coal and their mixtures. The two models are random hole model (RPM) and volume model (VM), respectively.

A Random hole model

The random pore model was put forward by Bhatia and Perlmutter in 1980, which considered the pore structure of reactive particles and their evolution in the reaction process. The model assumed that most of the pores on the particle surface were circular columnar pores with different pore diameters, and the chemical reaction between particles and gas phase occurred on these pore surfaces, and the reaction rate depended on the effective contact area, that is, the stacking degree of the pore surfaces. The dynamic equation is described as follows:

$$\frac{d\alpha}{dt} = k_{RPM}(1 - \alpha) \sqrt{1 - \psi \ln(1 - \alpha)} \tag{4-6}$$

where, k_{RPM} is reaction rate constant of KRPM random hole model, s^{-1}; ψ is the structural size of the reaction particles, the expression is $\psi = 4\pi L_0(1 - \varepsilon_0)/S_0^2 \cdot L_0$, S_0 and ε_0 represent the pore length per unit volume, the specific surface area of reaction and the porosity of particles when $t = 0$, respectively.

B Volume model

The model is relatively simple, assuming that the size of reaction particles does not change with the progress of the reaction, but the density of reactants changes linearly, that is, the combustion reaction rate has nothing to do with the particle size, and assuming that the reaction is a first-order reaction, its kinetic equation is described as follows:

$$\frac{d\alpha}{dt} = k_{VM}(1-\alpha) \tag{4-7}$$

where, k_{VM} is reaction rate constant of volume model, s^{-1}.

Combine Eq. (4-3) and Eq. (4-4) with Eq. (3-3) and Eq. (3-4) to obtain:

$$\alpha = 1 - \exp\left\{-\left[A_0 \frac{T-T_0}{\beta}\exp\left(\frac{-E}{RT}\right)\right]\left[1 + A_1 \frac{T-T_0}{4\beta}\exp\left(\frac{-E}{RT}\right)\right]\right\} \tag{4-8}$$

$$\alpha = 1 - \exp\left[-k_0 \frac{T-T_0}{\beta}\exp\left(\frac{-E}{RT}\right)\right] \tag{4-9}$$

where, $A_0 = \dfrac{k_0 C^n S_0}{1-\varepsilon_0}$, $A_1 = \dfrac{4\pi L_0 k_0 C^n}{S_0}$.

According to Eq. (4-5) and Eq. (4-6), the mathematical relationship between α and T obtained from the experiment is fitted nonlinearly, so as to solve the kinetic parameters k_0, E and ψ, and the calculated conversion rate of combustion reaction is compared with the experimental data to verify the accuracy of the two kinetic models in describing the kinetic process of mixed combustion of palm shell semi-coke and pulverized coal. At the same time, Eq. (4-7) is used to calculate the error between all calculated conversion values and experimental values obtained by different kinetic models, so as to measure the accuracy of the two kinetic models:

$$DEV(\alpha) = 100 \times \frac{\left[\sum_{i=1}^{N}(\alpha_{exp,i} - \alpha_{cale,i})^2/N\right]^{1/2}}{\max(\alpha)_{exp}} \tag{4-10}$$

where, $DEV(\alpha)$ is the relative error, %; $\alpha_{exp,i}$ is the experimental value of conversion rate; $\alpha_{cale,i}$ is the calculated value of conversion rate; $\max(\alpha)_{exp}$ is the maximum conversion rate, and the general value is about 1; N is the number of experimental points.

4.3.3.2 Combustion dynamics fitting results

The combustion kinetic parameters E, k_0 and ψ of the sample calculated according to the random hole model and the volume model are shown in Table 4-10, and the correlation coefficient R^2 of the fitting results is greater than 0.9969, which shows that the fitting effects of the two models are good. However, the fitting result of random hole model is slightly better than that of volume model, because the correlation coefficient R^2 of random hole model fitting is greater than that of volume model. According to the fitting results of random pore model, the activation energy of the sample is between 90.2kJ/mol and 121.8kJ/mol. With the increase of the proportion of palm shell semi-coke, the activation energy of the mixture first decreases and then increases, reaching the minimum value of 90.2kJ/mol when the amount of palm shell semi-coke is 60%, which indicates that the activation energy of the mixing process decreases after adding palm shell semi-coke to the pulverized coal.

4.3 Study on combustion characteristics of biomass char/pulverized coal in case of mixed injection

Table 4-10 Combustion Dynamics Parameters Calculated by RPM and VM Models

Proportion of palm shell	RPM				VM		
	E(kJ/mol)	k_0 (s^{-1})	ψ	R^2	E(kJ/mol)	k_0 (s^{-1})	R^2
0	116.3	2,791.5	6.91	0.998	143.3	263,717.0	0.99
20	121.8	15,500.3	5.09×10^{-15}	0.999	121.8	15,500.3	0.99
40	94.4	403.6	7.51×10^{-16}	0.998	94.4	403.6	0.99
60	90.2	361.0	2.14×10^{-21}	0.996	90.2	361.0	0.99
80	102.5	4,471.8	2.75×10^{-14}	0.997	102.5	4,471.8	0.99
100	113.3	40,613.0	0.21	0.998	116.6	74,379.6	0.99

The comparison between the activation energy calculated by the kinetic model fitting and the weighted average activation energy is shown in Fig. 4-21. The dotted line in the Fig. 4-21 is the weighted average activation energy of palm shell semi-coke and pulverized coal. It can be seen that the activation energy of the mixture is lower than that of palm shell semi-coke and pulverized coal burning alone. With the increase of the dosage of palm shell semi-coke, the activation energy of the mixture does not show linear additivity, which also proves that there is a synergistic effect in the mixed combustion process of palm shell semi-coke and pulverized coal. Barbanera et al.[32] found that the activation energy of the mixed combustion process of solid waste and charcoal was also not additive, and Wang et al.[18] also found the same change law when studying the mixed combustion dynamics of matter and low-rank pulverized coal.

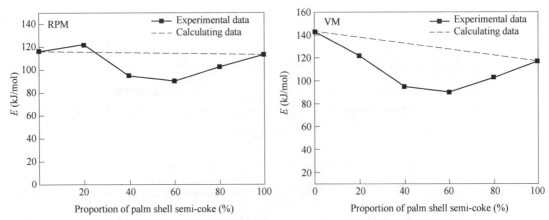

Fig. 4-21 Combustion Dynamics Parameters Calculated by RPM and VM Models

The two models make different assumptions when describing the reaction process, which will introduce a series of errors in data processing. In order to quantitatively compare the advantages and disadvantages of the two models in describing the combustion process of palm shell semi-coke, pulverized coal and their mixture, the most suitable kinetic model is selected, and the fitted kinetic parameters are brought into Eq. (4-8) and Eq. (4-9), respectively. The relationship between the calculated reaction conversion and the reaction temperature at four heating rates is compared and analyzed with the actual curve measured by experiments, the results are shown in Fig. 4-22. It can be clearly seen that the calculated curves of the random pore model and the volume

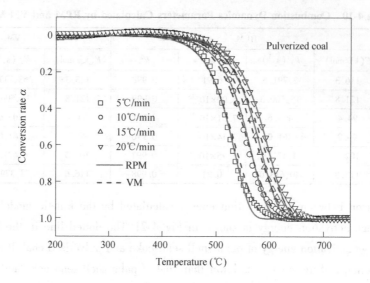

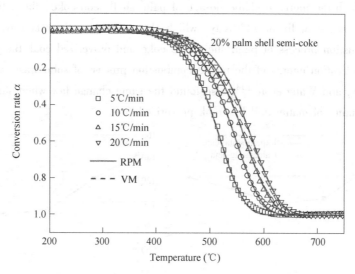

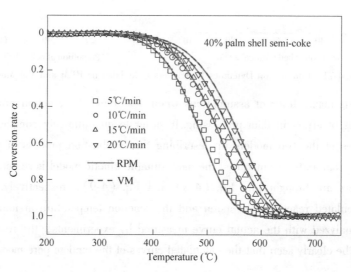

4.3 Study on combustion characteristics of biomass char/pulverized coal in case of mixed injection

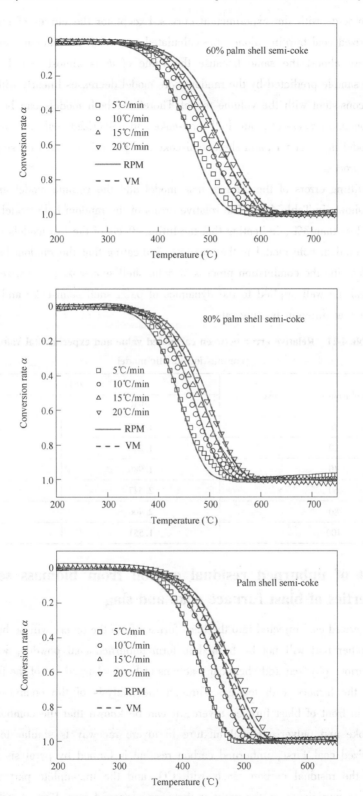

Fig. 4-22 Comparison of experimental values and calculated values of different models

model almost coincide with the experimental curves. Except for the curves of brown shell semi-coke and pulverized coal burning alone, the calculated results of the random pore model and the volume model are almost the same because the value of ψ is almost zero. In this case, the reactivity of the sample predicted by the random pore model decreases linearly with the conversion rate, which is consistent with the volume model. Therefore, both models can be used to predict the mixed combustion process of palm shell semi-coke and pulverized coal, but the fitting effect of random hole model is better for palm shell semi-coke and pulverized coal with single substance in the combustion process.

The specific fitting errors of the random hole model and the volume model are calculated by Eq. (4-7), as shown in Table 4-11. The relative errors of the random hole model and the volume model are both less than 3%, indicating that the fitting effects of the two models are better, while the error of the random hole model is the smallest, indicating that the random hole model is the best model to describe the combustion process of palm shell semi-coke, pulverized coal and their mixtures, and can be well applied to the dynamics of palm shell semi-coke and pulverized coal and their mixtures at different heating rates.

Table 4-11 Relative error between calculated value and experimental value of combustion kinetic model (%)

Proportion of palm shell semi-coke	$DEV(\alpha)$	
	RPM	VM
0	1.820	2.066
20	1.260	1.260
40	1.986	1.986
60	2.547	2.547
80	2.406	2.406
100	1.551	1.569

4.4 Effect of unburned residual carbon from biomass semi-coke on properties of blast furnace coke and slag

Most of the pulverized coal injected into the blast furnace from the tuyere will be burned before the tuyere, and another part will not be burned to form unburned coal powder, which will either participate in various physical and chemical reactions or be discharged out of the furnace with slag or escape from the furnace with top gas. Through the analysis of the combustion behavior of pulverized coal in front of blast furnace tuyere, it can be known that the combustion process of biomass semi-coke and pulverized coal mixture in tuyere raceway is similar to that of simply injecting pulverized coal. First, unburned carbon residue is formed by pyrolysis and combustion reaction. Then, the residual carbon reacts with CO_2 and the incomplete part enters the blast furnace column to participate in the reaction between slag and iron. Part of unburned residual carbon will adhere to the surface of coke, which may have a certain impact on the carbon melting

— 4.4 Effect of unburned residual carbon from biomass semi-coke on properties of blast furnace coke and slag

loss reaction of coke, thus weakening the role of coke as the framework of blast furnace; part of unburned carbon residue will enter the slag and deposit in the slag, which is an important consumption way of unburned residual carbon in blast furnace. The existence of unburned residual carbon will affect the viscosity and other properties of slag. Different combustion rates and ash compositions of biomass semi-coke and pulverized coal will lead to different unburned residual carbon in tuyere raceway. Therefore, it is necessary to study the influence of unburned residual carbon produced after mixed injection of semi-coke and pulverized coal on gasification reaction of blast furnace coke and slag viscosity.

4.4.1 Effect of unburned residual carbon on coke gasification reaction

As an indispensable fuel in blast furnace ironmaking, coke plays a very important role in blast furnace ironmaking process. The functions of coke in blast furnace are mainly divided into: (1) reductant;(2) heating agent; (3) carburizing agent; (4) skeleton action, among which skeleton action is the most important and cannot be replaced by other substances. Under the condition of high coal injection ratio, the skeleton function of coke is more prominent, and the quality requirements of coke are higher and higher. In this section, palm shell semi-coke prepared by 600℃ pyrolysis used in the mixed combustion of Section 4.3 was selected, and the co-gasification reaction of unburned residual carbon and coke was studied by thermogravimetric analysis, and the influence of unburned residual carbon on coke gasification reaction was analyzed.

4.4.1.1 Experimental equipment and method

At present, unburned coal powder cannot be obtained from blast furnace, so the experimental equipment used to prepare unburned residual carbon under laboratory conditions is ZK-16XQ-1700 ventilated muffle furnace produced by Beijing Zhongke Beiyi Technology Co., Ltd., as shown in Fig. 4-23.

In order to compare and analyze the influence of palm shell semi-coke injection on coke in blast furnace, two groups of unburned residual carbon were prepared, one was Yangquan anthracite, the other was a mixture of palm shell semi-coke and Yangquan anthracite, and the proportion of palm shell semi-coke was 20%. The preparation process was to weigh 5g pulverized coal and the mixture of palm shell semi-coke and pulverized coal, respectively, spread them evenly in alumina porcelain boats, put them in muffle furnace, and set up the heating program. Nitrogen was introduced at the flow rate of 1L/min, and the air in the muffle furnace was exhausted first, and then the temperature was raised. Nitrogen was introduced for protection during the whole experiment. Two groups of samples were heated to 1,200℃ with the furnace, and dry distilled at 1,200℃ for 1h to remove the volatiles in the pulverized coal, then the heating was stopped, and the samples were cooled to room temperature with the furnace and taken out to obtain unburned residual carbon. Two kinds of unburned residual carbon are defined as No. 1 residual carbon and No. 2 residual carbon, respectively.

Fig. 4-23 Equipment for prepare unburned residual carbon

The coke selected in the experiment is the coke used in the blast furnace of a steel plant, and its composition analysis is shown in Table 4-12. The coke is ground and sieved, and coke particles with particle size below 0.074mm are selected for standby. In order to study the influence of unburned residual carbon on the gasification reaction of coke, two kinds of unburned residual carbon are mixed with coke at a mass fraction of 10%, respectively, and ground with agate mortar for 10min until they are evenly mixed. The gasification reaction experiments of coke, residual carbon and the mixture of coke and residual carbon were carried out on the HCT-3 thermogravimetric analyzer. The experimental system automatically and continuously sampled, recorded the experimental data and drew the weight loss curve. In each experiment, 5mg of samples were weighed with alumina crucible and put on the differential thermal balance, and the temperature was raised from room temperature to 1,300℃ at a heating rate of 5℃/min. During the experiment, CO_2 was introduced at a flow rate of 60mL/min, and the experimental data of each group were recorded. The conversion rate α in the gasification reaction of the sample is calculated from the data recorded in the weight loss curve, and its definition is shown in Eq. (3-1).

Table 4-12 Proximate analysis of coke and unburned carbon residue (%)

Sample	M_{ad}	V_{ad}	A_{ad}	FC_{ad}
Coke	0.74	1.50	11.14	86.62
Pulverized coal	0.84	7.90	12.32	78.94
No. 1 residual carbon	0.92	1.25	13.59	84.24
No. 2 residual carbon	0.84	1.27	13.47	84.42

4.4.1.2 Sample performance analysis

See Table 4-12 for the composition analysis of coke and unburned residual carbon. The chemical composition contents of coke and two kinds of residual carbon are very close, and the volatile content of coke is very low. During the dry distillation process, the volatile content of pulverized coal, palm shell semi-coke and the mixture of pulverized coal is gradually precipitated under nitrogen atmosphere, and the volatile content of unburned residual carbon obtained after dry distillation treatment is obviously reduced, while the content of ash and fixed carbon is increased.

By scanning electron microscope, the micro-morphology of pulverized coal and unburned residual carbon obtained after high-temperature dry distillation treatment is compared. Fig. 4-24 shows the micro-morphology of raw coal and unburned residual carbon enlarged by 2,000 times. As can be seen from the Fig. 4-24, raw coal and unburned residual carbon mainly exist in the form of particles, and the surface of raw coal is smooth, the structure is relatively compact, and there are no obvious pores and cracks. However, the unburned residual carbon after high-temperature dry distillation will form some cracks on the surface due to the escape of volatiles during the treatment, and the surface is not as smooth as the raw coal, which becomes rough, and some fine particles or debris are attached.

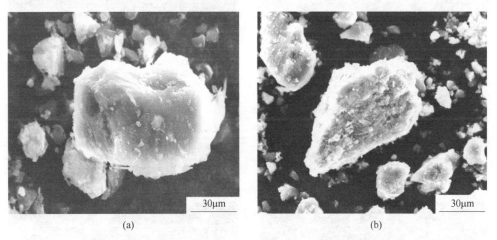

Fig. 4-24 Comparison of electron microscope pictures of raw coal and unburned residual carbon
(a) pulverized coal; (b) unburned residual carbon

Fig. 4-25 shows the microscopic morphology comparison of coke and two kinds of unburned residual carbon at magnification of 500 times and 3,000 times, respectively. As can be seen from the Fig. 4-25, both coke and unburned residual carbon exist in the form of particles, and the surface of coke is relatively rough, and some small fragmented substances are attached. There are some fine particles and some spherical particles attached to the surface of No. 1 residual carbon, but there are no obvious pores and cracks on the surface. No. 2 residual carbon is obtained by dry distillation of a mixture of palm shell semi-coke and pulverized coal. It can be seen that there are obviously two kinds of particles in No. 2 residual carbon, one is the particles obtained by dry distillation of pulverized coal. The other is the particles obtained by carbonization of palm shell semi-coke, and there are many obvious holes on the surface. Palm shell semi-coke is made by pyrolysis of palm shell, and its pores are developed. After dry distillation, some small pores of palm shell semi-coke particles fuse to form larger pores, and many spherical particles similar to No. 1 residual carbon are attached to the surface.

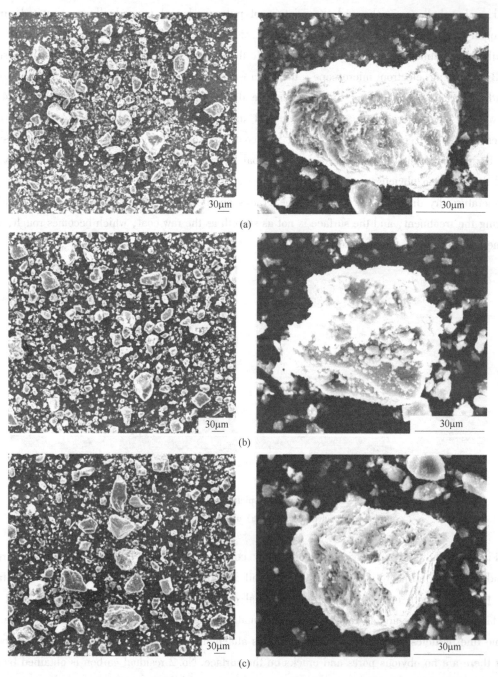

Fig. 4-25 Comparison of electron microscope pictures of coke and unburned residual carbon
(a) coke; (b) No. 1 unburned residual carbon; (c) No. 2 unburned residual carbon

The spherical particles attached to the surface of carbon residue were analyzed by energy spectrum, and the results are shown in Fig. 4-26, which shows that the main elements of spherical particles are C, Si, Ca and O, in which the mass fraction of C is 61.35%, the mass fraction of Si is 4.54%, the mass fraction of Ca is 3.67% and the mass fraction of O is 8.05%. It can be considered that spherical particles are precipitated and attached to residual carbon particles after high temperature treatment.

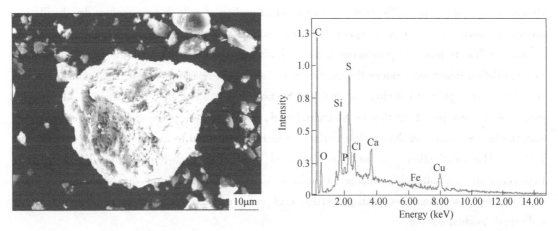

Fig. 4-26 Energy spectrum analysis results of spherical particles on the surface of unburned residual carbon

The carbon structure has always been considered as an important factor affecting the gasification reactivity of coke[33], so the samples of coke, No. 1 residual carbon and No. 2 residual carbon were analyzed by X-ray diffraction. As shown in Fig. 4-27, the (002) peak of coke was narrower than that of the two kinds of residual carbons. Through calculation, it is found that the spacing d_{002} of microcrystal lamellae of coke, No. 1 residual carbon and No. 2 residual carbon is 0.345nm, 0.753nm and 0.770nm, respectively, and the average stacking height L_c of microcrystal lamellae

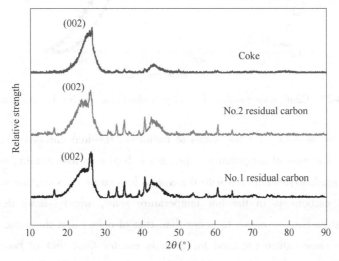

Fig. 4-27 X-ray diffraction patterns of coke, No. 1 residual carbon and No. 2 residual carbon

is 2.717nm, 2.380nm and 2.203nm, respectively. Therefore, the carbon order of coke is greater than that of two kinds of residual carbons, and that of No. 1 residual carbon is greater than that of No. 2 residual carbon.

4.4.1.3 Effect of unburned residual carbon on gasification characteristics of coke

Compare and analyze the gasification reaction curves of raw coal and No. 1 residual carbon, as shown in Fig. 4-28. The gasification curve of residual carbon after high-temperature dry distillation obviously moves to the high-temperature area, which shows that the gasification reaction of residual carbon is less likely to occur compared with pulverized coal, and the high-temperature dry distillation treatment reduces the gasification reactivity of residual carbon. The results obtained from the microscopic morphology analysis in Section 4.4.1 show that the structure of residual carbon is less compact than that of pulverized coal, and with the escape of volatiles, a few pores and cracks are formed on the particle surface, which is beneficial to the contact between CO_2 and particles. The gasification reaction was promoted, but after high-temperature dry distillation treatment, the graphitization degree of residual carbon increased and the amorphous carbon structure decreased compared with pulverized coal, which led to the decrease of the reactivity of unburned residual carbon.

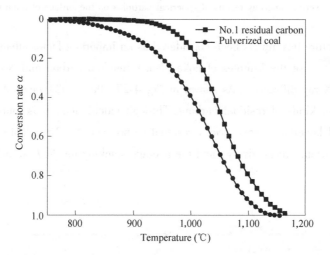

Fig. 4-28 Gasification reaction curve of pulverized coal and No. 1 residual carbon

The reaction curve of coke with two kinds of unburned residual carbons and CO_2 is shown in Fig. 4-29. With the increase of temperature, the sample begins to lose weight, and the gasification reaction proceeds gradually. Compared with the weight loss curve of coke, the weight loss curve of unburned residual carbons is in the low temperature zone, which shows that the gasification reaction of residual carbons is easier to occur than that of coke, and the same conversion rate is achieved, and the temperature required for coke is greater than that of two kinds of residual carbons. Comparing the two kinds of residual carbons, it is found that the gasification weight loss curve of No. 2 residual carbon moves to the low temperature region compared with the gasification reaction weight loss curve of No. 1 residual carbon, and the gasification reaction of No. 2 residual carbon is easier to occur than that of No. 1 residual carbon. This may be caused by three reasons: First, the difference of carbon structure, X-ray diffraction analysis results show that the carbon order of No. 1 residual carbon is greater than that of No. 2 residual carbon. The second is the difference of microstructure. The results of microstructure analysis show that there are a lot of pore structures on the surface of No. 2 residual carbon particles, which are more developed than No. 1 residual carbon particles, which is beneficial to contact with CO_2 during the reaction. Thirdly, the alkali metal and alkaline earth metal oxides brought in from brown shell semi-coke can catalyze the gasification reaction of No. 2 residual carbon.

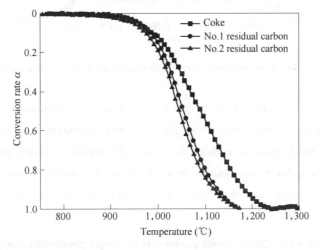

Fig. 4-29 Gasification reaction curve of coke and residual carbon

Fig. 4-30 shows the comparative analysis of the gasification weight loss curve of coke and residual carbon mixture and coke. As can be seen from the Fig. 4-30, compared with the weight loss curve of coke, the weight loss curve of the mixture of coke and residual carbon moves to the low temperature region, which shows that the gasification reaction of coke can occur at a lower temperature after adding unburned residual carbon, and the addition of residual carbon catalyzes the gasification reaction of coke. However, there is little difference among the three weight loss curves, indicating that although the addition of residual carbon catalyzes the gasification reaction of coke, the catalytic effect is not obvious.

The characteristic parameters of each gasification reaction experiment are calculated by the method introduced in Section 4. 3. 2. The results are shown in Table 4-13 and Fig. 4-31. The characteristic parameters of gasification reaction can directly and concretely characterize the gasification reaction characteristics of the sample. From the calculated data, it can be seen that the gasification reaction end temperature T_f and the maximum weight loss rate temperature T_m of coke are both higher than those of two kinds of residual carbons, but the reactivity index $R_{0.5}$, comprehensive reaction characteristic index S and ignition index C of coke are all lower than those of two kinds of residual carbons, which shows that. At the same time, comparing the two kinds of

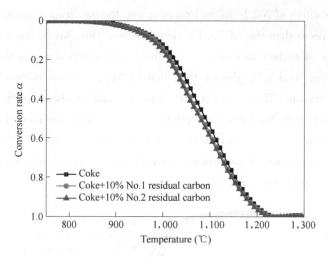

Fig. 4-30 Reaction curve of co-gasification of coke and residual carbon

residual carbons, it can be found that the gasification reaction start temperature T_i, gasification reaction end temperature T_f and maximum weight loss rate temperature T_m of No. 2 residual carbon are all lower than No. 1 residual carbon. However, the reactivity index $R_{0.5}$, comprehensive reaction characteristic index S and ignition index C of No. 2 residual carbon are all greater than No. 1 residual carbon. These data show that the gasification reactivity of No. 2 residual carbon is greater than No. 1 residual carbon.

Table 4-13 Characteristic parameters of sample gasification reaction

Sample	T_i (℃)	T_m (℃)	T_f (℃)	$R_{0.5} \times 10^{-5}$ (s^{-1})	$S \times 10^{-16}$ (s$^{-2} \cdot ℃^{-3}$)	$C \times 10^{-10}$ (s$^{-1} \cdot ℃^{-2}$)
No. 1 residual carbon	961	1,051	1,158	3.95	1.36	7.79
No. 1 residual carbon	951	1,049	1,151	3.99	1.38	7.99
Coke	953	1,093	1,217	3.82	1.08	4.51
Coke+10% No. 1 residual carbon	948	1,078	1,214	3.85	1.11	4.58
Coke+10% No. 2 residual carbon	940	1,072	1,209	3.86	1.16	4.70

Comparing the gasification characteristic parameters of coke and mixed samples with two kinds of residual carbons, respectively, the gasification reaction start temperature T_i, gasification reaction end temperature T_f and maximum weight loss rate temperature T_m of coke+No. 1 residual carbon and coke+No. 2 residual carbon are all lower than coke, while the reactivity index $R_{0.5}$, comprehensive reaction characteristic index parameter S and ignition index C of coke + No. 1 residual carbon and coke+No. 2 residual carbon are higher than coke. These data show that the gasification reactivity of coke and residual carbon mixture is better than coke. Comparing the

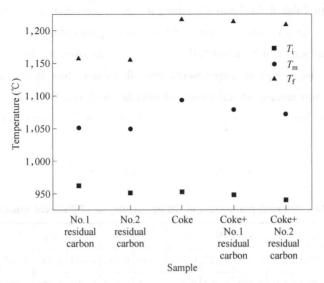

Fig. 4-31　Comparison of T_i, T_m and T_f of different samples

characteristic parameters of coke + No. 1 residual carbon and coke + No. 2 residual carbon, the gasification reaction starting temperature T_i of coke + No. 1 residual carbon is 948℃, which is higher than that of coke+No. 2 residual carbon. The gasification reaction end temperature T_f and the gasification maximum temperature T_m of coke+No. 1 residual carbon and coke+No. 2 residual carbon are 1,214℃, 1,209℃ and 1,078℃, 1,072℃, respectively. The sample of coke+No. 1 residual carbon is slightly lower than that of coke+No. 2 residual carbon. At the same time, the reactivity index $R_{0.5}$, comprehensive reaction characteristic index S and ignition index C of coke+ No. 1 residual carbons are all smaller than those of coke+No. 2 residual carbon, which shows that the catalytic effect of No. 2 residual carbon on coke gasification reaction is greater than that of No. 1 residual carbon. As can be seen from Fig. 4-31, the gasification characteristic parameters of coke with two kinds of unburned residual carbons do not change much compared with coke, which shows that the catalytic effect of unburned residual carbons on coke gasification reaction is not obvious, and the difference between the characteristic parameters of two kinds of unburned residual carbons is not big, which shows that the catalytic effect of unburned residual carbons on coke gasification reaction is strengthened after adding palm shell semi-coke. However, the enhancement is not large.

4.4.1.4　Effect of unburned residual carbon on coke gasification kinetics

According to the method of determining the reaction mechanism function and calculating the kinetic parameters introduced in Section 4.3.3, the kinetic analysis of the mixed gasification reaction of coke and unburned residual carbon is carried out, and the mechanism function and kinetic parameters that accurately describe the dynamic process of coke gasification reaction are solved.

The correlation coefficients calculated by different mechanism functions obtained through fitting analysis are shown in Table 4-14. From the table, it can be seen that $A_{3/2}$ model is the best model to describe the coke gasification reaction, and the coke gasification reaction after adding two unburned residual carbons can be described by A_1 model, but overall, the correlation coefficients of $A_{3/2}$ model in three groups of experiments are all greater than 0.99, and the correlation coefficient is the largest among several groups of models, so it can be considered that $A_{3/2}$ model is the best model to describe the coke gasification reaction in this experiment, and its reaction mechanism function is:

$$f(\alpha) = \frac{3}{2}(1-\alpha)[-\ln(1-\alpha)]^{\frac{1}{3}} \qquad (4\text{-}11)$$

Table 4-14　Relevant parameters of fitting calculation of different kinetic models

Sample	A_1	A_2	A_3	A_4	$A_{3/2}$	O_2	O_3	R_2	R_3	D_1	D_2	D_3	D_4	MaxR²	Model
No. 1	0.952	0.995	0.993	0.991	0.996	0.771	0.794	0.976	0.985	0.949	0.968	0.987	0.975	0.996	$A_{3/2}$
No. 2	0.993	0.991	0.987	0.983	0.992	0.782	0.805	0.968	0.979	0.937	0.959	0.981	0.968	0.993	A_1
No. 3	0.994	0.992	0.990	0.985	0.993	0.770	0.794	0.972	0.981	0.943	0.963	0.984	0.971	0.994	A_1

Note: No. 1 is coke, No. 2 is coke+10% No. 1 residual carbon, and No. 3 is coke+10% No. 2 residual carbon.

The mechanism of this model is random nucleation and subsequent growth (Aveami-Erofeev equation). Through the kinetic method introduced in Section 4.3.3, the kinetic parameters of each gasification reaction calculated by the above model are shown in Table 4-15, and the activation energy of coke gasification reaction is 140.1kJ/mol. The activation energy of coke gasification reaction with 10% No. 1 residual carbon and No. 2 residual carbon is 137.3kJ/mol and 133.8kJ/mol, respectively. Adding unburned residual carbon reduces the activation energy of coke gasification reaction, which shows that unburned residual carbon catalyzes the coke gasification reaction, and the catalytic effect of No. 2 residual carbon is slightly greater than that of No. 1 residual carbon, which is consistent with the conclusion in the previous section.

Table 4-15　Kinetic parameters of sample gasification reaction

Sample	$\ln(A)$ (s⁻¹)	E(kJ/mol)
No. 1	4.9	140.1
No. 2	4.7	137.3
No. 3	4.4	133.8

Note: No. 1 is coke, No. 2 is coke+10% No. 1 residual carbon, and No. 3 is coke+10% No. 2 residual carbon.

Through the above calculation and analysis, the activation energy, pre-exponential factor and optimal mechanism function of coke gasification reaction in this experiment are obtained, so that the dynamic model describing coke gasification reaction can be determined. The dynamic process of coke gasification reaction after adding two unburned residual carbon is described as follows:

$$\frac{d\alpha}{dt} = 2.233 \times 10^3 \exp\left(-\frac{140,079.4}{RT}\right)(1-\alpha)[-\ln(1-\alpha)]^{\frac{1}{3}} \qquad (4\text{-}12)$$

4.4 Effect of unburned residual carbon from biomass semi-coke on properties of blast furnace coke and slag

$$\frac{d\alpha}{dt} = 1.914 \times 10^3 \exp\left(-\frac{137,319.1}{RT}\right)(1-\alpha)[-\ln(1-\alpha)]^{\frac{1}{3}} \quad (4\text{-}13)$$

$$\frac{d\alpha}{dt} = 1.430 \times 10^3 \exp\left(-\frac{133,776.0}{RT}\right)(1-\alpha)[-\ln(1-\alpha)]^{\frac{1}{3}} \quad (4\text{-}14)$$

Through the above analysis, it can be concluded that the gasification reactivity of unburned residual carbon is greater than that of coke, palm shell semi-coke enhances the gasification reactivity of unburned residual carbon, and unburned residual carbon adheres to the surface of coke and will preferentially react with CO_2, which can protect coke to some extent. At the same time, the catalytic effect of unburned residual carbon on coke gasification reaction is not obvious, which will not cause further deterioration of coke quality.

4.4.2 Effect of unburned residual carbon on viscosity of blast furnace slag

The unburned coal particles generated from coal injection in blast furnace rise with the gas flow and some of them enter the slag, which changes the viscosity and performance of the slag, thereby affecting blast furnace smelting. Therefore, this section mainly focuses on the impact of unburned residual coke generated from palm kernel shell char injection on the viscosity of slag.

4.4.2.1 Experimental equipment and methods

Under experimental conditions, a quaternary system slag was prepared using analytical grade reagents CaO (purity ≥ 98.0%), SiO_2 (purity ≥ 99.0%), MgO (purity ≥ 98.5%), and Al_2O_3 (purity ≥ 99.0%) to maintain the alkalinity and the content of the four oxides in the slag unchanged. Different masses and compositions of unburned residual coke were added to study their impact on the viscosity of the quaternary system slag. The specific composition and proportional scheme of the prepared slag are presented in Table 4-16 and Table 4-17.

Table 4-16 Chemical composition and proportion scheme of slag

Number	CaO	SiO_2	MgO	Al_2O_3	Slag	No. 1 residual carbon	No. 2 residual carbon
S0	40.86	37.14	8.0	14.0	100.0	—	—
S1	40.86	37.14	8.0	14.0	99.5	0.5	—
S2	40.86	37.14	8.0	14.0	99.0	1.0	—
S3	40.86	37.14	8.0	14.0	98.0	2.0	—
S4	40.86	37.14	8.0	14.0	99.0	—	1.0

Table 4-17 Dosage of each experimental sample

Number	CaO	SiO_2	MgO	Al_2O_3	No. 1 residual carbon	No. 2 residual carbon	Total mass
S0	57.20	52.00	11.20	19.60	—	—	140.0
S1	56.92	51.74	11.14	19.50	0.7	—	140.0
S2	56.63	51.48	11.09	19.40	1.4	—	140.0

Continued Table 4-17

Number	CaO	SiO₂	MgO	Al₂O₃	No. 1 residual carbon	No. 2 residual carbon	Total mass
S3	56.06	60.96	10.98	19.21	2.8	—	140.0
S4	56.63	51.48	11.09	19.40	—	1.4	140.0

The viscosity of the test samples in this experiment was measured using the rotational cylinder method, with the RTW-10 melt property comprehensive measuring instrument produced by Northeastern University as the viscosity apparatus. The equipment mainly consists of a computer, a viscosity meter, a control cabinet, and a high-temperature resistance furnace with a U-shaped $MoSiO_2$ heating rod, as shown in Fig. 4-32. The internal diameter of the alumina furnace tube is 53mm, and the temperature inside the furnace tube is directly controlled by a computer program, using two platinum-rhodium thermocouples (Pt-10%Rh/Pt) for temperature measurements. Prior to the measurement experiment, the viscosity constant was calibrated with castor oil as an analytical reagent at room temperature. The weighed samples were mixed uniformly and loaded into graphite crucibles with a diameter of 39mm and a height of 60mm, which were then placed in the constant temperature zone of the furnace. The program was set to increase the temperature inside the furnace at a rate of 10℃/min to 1,000℃, and then at a rate of 5℃/min to 1,520℃, followed by a 60-minute holding time to ensure uniform melting of the slag sample. The entire reaction process was protected by high-purity argon gas (purity ≥ 99.999%, flow rate of 1L/min) and cooled with water. After reaching the predetermined temperature, the viscosity of the slag was measured at different rotational speeds (100r/min, 150r/min and 200r/min) and different temperatures (1,520℃, 1,510℃, 1,500℃, 1,490℃, 1,480℃ and 1,460℃) using the constant temperature viscosity measurement mode.

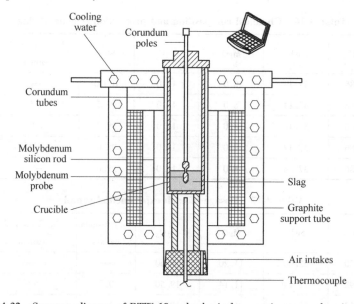

Fig. 4-32 Structure diagram of RTW-10 melt physical properties comprehensive tester

4.4.2.2 The effect of temperature on slag viscosity

Experimental measurements of viscosity values for various groups of samples under different temperature and rotational speed conditions are presented in Table 4-18. From the data in the table, it can be concluded that the S0 group samples, which were not mixed with unburned residual carbon furnace slag, belong to a homogeneous melt and their viscosity values do not vary significantly with rotational speed. The other furnace slag samples, mixed with unburned residual carbon, belong to heterogeneous bodies, and their viscosity values exhibit changes with both temperature and rotational speed.

Table 4-18 The viscosity values of experimental samples measured under different temperature and rotational speed conditions (dPa·s)

Sample	Speed (r/min)	Temperature (℃)					
		1,520	1,510	1,500	1,490	1,480	1,460
S0	200	2.5854	2.7643	2.9148	3.0514	3.2948	3.6951
	150	2.5865	2.7658	2.9136	3.0586	3.2936	3.6983
	100	2.5887	2.7641	2.9128	3.0591	3.2933	3.6968
S1	200	2.8692	3.0245	3.1762	3.3753	3.5385	4.0299
	150	2.9583	3.1344	3.2984	3.4704	3.6277	4.0794
	100	3.0782	3.2286	3.3944	3.5493	3.7083	4.1235
S2	200	3.3332	3.5885	3.8571	4.1766	4.5134	5.3197
	150	3.4562	3.7288	4.0448	4.3945	4.7555	5.6839
	100	3.6983	4.0184	4.3825	4.7558	5.1787	6.0935
S3	200	6.9216	7.2818	7.7369	8.1205	8.5513	9.2843
	150	7.5892	7.992	8.3975	8.7818	9.1388	9.7837
	100	8.4531	8.7913	9.1463	9.4896	9.8506	10.4247
S4	200	3.2382	3.5315	3.8023	4.0741	4.4595	5.0581
	150	3.3821	3.6784	3.9916	4.2896	4.6611	5.3048
	100	3.6587	3.9373	4.2489	4.5828	4.8987	5.5513

The Arrhenius[34] study revealed an exponential relationship between the viscosity of the melt and temperature, which can be described by Eq. (4-15).

$$\eta = A\exp\left(\frac{E_\eta}{RT}\right) \qquad (4-15)$$

where η ——Viscosity, Pa·s;
 A ——Pre exponential factor;
 E_η ——Viscous flow activation energy, J/mol;
 T ——Absolute temperature, K;
 R ——Gas constant, J/(mol·K).

$$\ln\eta = \frac{E_\eta}{R} \cdot \frac{1}{T} + \ln A \qquad (4\text{-}16)$$

Based on the viscosity measurements shown in Table 4-18, the logarithm of viscosity values ($\ln \eta$) for various groups of slag samples at different conditions were plotted against the reciprocal of temperature ($1/T$), as shown in Fig. 4-33. It can be observed that $\ln\eta$-$1/T$ for all five groups of slag samples are almost linear under different rotating speeds. This indicates that the viscosity-temperature relationship for these slag samples at different rotating speeds conforms to the Arrhenius equation, thereby suggesting the validity of this relationship for both homogeneous and heterogeneous melts. Furthermore, it can be clearly deduced that when the rotating speed is constant, the viscosity of slag increases as temperature decreases.

Fig. 4-33 (a) demonstrates the variation of viscosity of homogeneous melt without unburned residual carbon as a function of temperature under different rotational speed conditions. The three lines depicted as $\ln\eta$-$1/T$ at 200r/min, 150r/min and 100r/min exhibit a remarkable overlapping behavior, indicating that the rotational speed has virtually no influence on the viscosity of the homogeneous melt. Therefore, the homogeneous melt can be classified as a Newtonian fluid.

Fig. 4-33 (b), (c) and (d) depict the variation of slag viscosity with temperature for different proportions of No.1 residual carbon added. When the mass fraction of the unburned residual carbon added is low (0.5%), the distance between the curves corresponding to the three rotational speeds is small, indicating a small difference in viscosity among different rotational speeds. As the proportion of unburned residual carbon additive increases, the distance between the three curves gradually increases, indicating a gradual increase in the difference in viscosity among different rotational speeds. The effect of rotational speed on slag viscosity becomes increasingly significant, and the slag gradually loses its Newtonian fluid behavior. As a result, the viscosity of homogeneous slag does not vary with rotational speed, whereas the viscosity of non-homogeneous slag varies with changes in rotational speed. At constant rotational speeds, slag viscosity increases as temperature decreases.

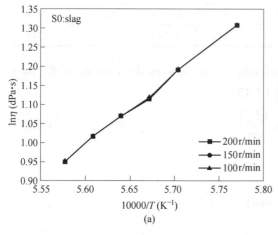
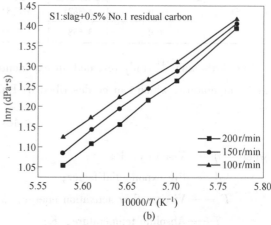

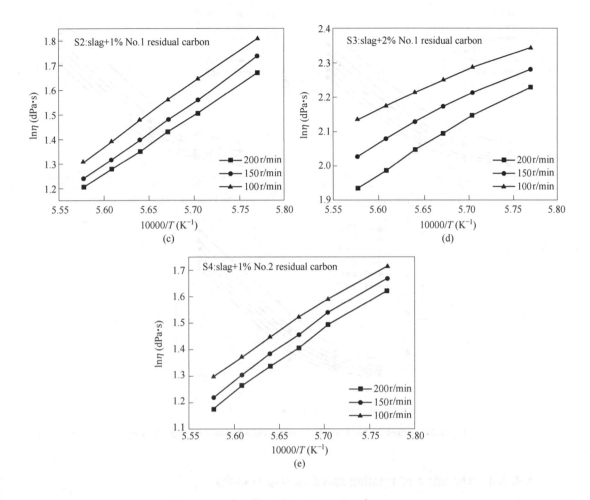

Fig. 4-33 The effect of temperature on slag viscosity

4.4.2.3 Effect of unburned residual carbon content on slag viscosity

The graph shown in Fig. 4-34 illustrates the variations in slag viscosity with respect to the mass fraction of residual coal (i.e., No. 1) added to the slag at different temperature conditions. Based on this graph, it can be observed that as the mass fraction of No. 1 residual coal increases, the slag viscosity exhibits an almost linear growth trend under all three different rotational speeds (i.e., 200r/min, 150r/min and 100r/min) conditions. At the same rotational speed but different temperatures, the trend of changes in slag viscosity with respect to the mass fraction of No. 1 residual coal is consistent, suggesting that the mass fraction of residual coal inside the slag is a significant factor affecting slag viscosity. Moreover, as the mass fraction of No. 1 residual coal increases, the magnitude of the increase in slag viscosity also increases. When the No. 1 residual coal content in the slag reaches 2%, the maximum slag viscosity exceeds 10dPa·s. Therefore, it is recommended to limit the content of residual coal in the slag.

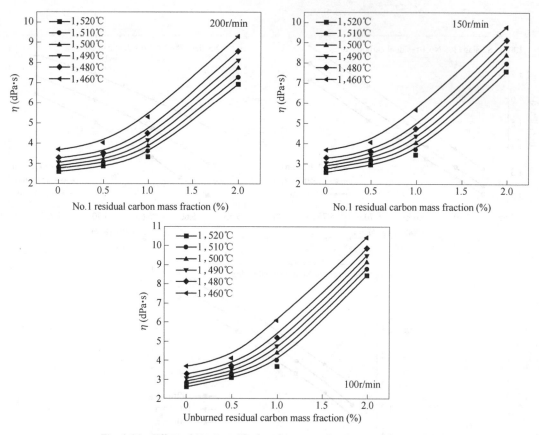

Fig. 4-34 Effect of No. 1 residual carbon mass fraction on slag viscosity

4.4.2.4 The effect of rotating speed on slag viscosity

The curve of the variation of slag viscosity with rotational speed is illustrated in Fig. 4-35. It can be observed that at the same temperature, the slag viscosity increases with the increase of residual carbon mass fraction. Under various temperature conditions (1,520℃, 1,510℃, 1,500℃, 1,490℃, 1,480℃ and 1,460℃) in this experiment, excluding the basic slag without added residual carbon, the viscosity of slag increases as rotational speed decreases, exhibiting the same phenomenon under different residual carbon mass fraction conditions. As rotational speed increases, the slag viscosity gradually decreases. This can be attributed to the fact that the slag with added residual carbon is a high-temperature solid-liquid mixed melt, which exhibits the property of shear-thinning with the increase of rotational speed. This indicates that rotational speed is also one of the important factors affecting the viscosity of non-homogeneous slag.

From Fig. 4-35, it can be observed that the variation in slag viscosity with respect to rotational speed is related to the mass fraction of unburned residual carbon. As the mass fraction of residual carbon of type 1 increases, the range of variation in slag viscosity with respect to rotational speed gradually intensifies. This indicates that the more residual carbon of type 1 present in the slag melt, the more prominent the influence of rotational speed on slag viscosity. Furthermore, under several mass fraction conditions of residual carbon of type 1 in this experiment, the range of variation in slag viscosity between 200-150r/min is smaller than that of 150-100r/min, as an increase in rotational speed leads to a more homogeneous distribution of unburned residual carbon in the slag.

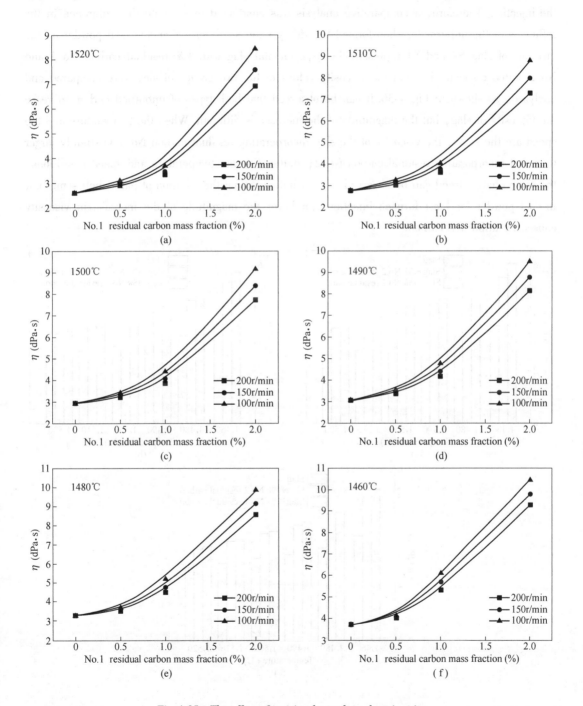

Fig. 4-35 The effect of rotational speed on slag viscosity

4.4.2.5 Effect of unburned residual carbon composition on slag viscosity

The mixture of palm shell semi-char and coal powder used in blast furnace injection will cause a change in the remaining unburned carbon components that leave the raceway swirling zone after the injection. Therefore, a comparative analysis was conducted to evaluate the difference in the influence of the unburned carbon formed by adding palm shell semi-char and coal powder on the viscosity of slag. S2 and S4 represent the experimental slag with 1% residual carbon No. 1 and No. 2 added respectively, and the viscosity values of the two groups of slag were compared and analyzed, as shown in Fig. 4-36. It can be observed that both types of unburned carbon increase the viscosity of slag, but the magnitude of the increase is different. When the temperature and the speed are the same, the viscosity of the slag incorporating residual carbon No. 1 is slightly larger than that incorporating residual carbon No. 2; under different temperature and speed conditions, the same change trend can be observed, which indicates that the addition of palm shell semi-char in coal powder for blast furnace injection can lower the magnitude of the increase in viscosity caused by unburned carbon.

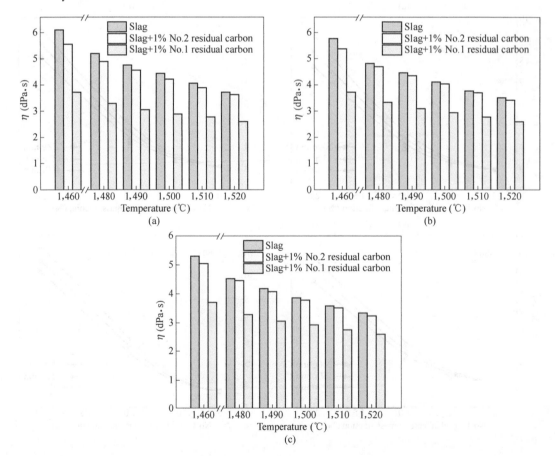

Fig. 4-36 Comparison of the effects of No. 1 residual carbon and No. 2 residual carbon on slag viscosity
(a) 100r/min; (b) 150r/min; (c) 200r/min

The two types of unburned residual char are both solid particles that exist in the form of solid particles in the slag, both of which cause an increase in the viscosity of the slag. When the same mass fraction is added, there is a difference in the viscosity of the two groups of slag. From the composition of the ash content of the coal and palm kernel shells, it is found that the main substances in the coal ash are SiO_2 and Al_2O_3, with contents of 49.61% and 37.30%, respectively, whereas the contents of SiO_2 and Al_2O_3 in the palm kernel shell ash are only 1.95% and 0.66%, respectively. The main substances in the palm kernel shell ash are CaO and MgO, with contents of 25.86% and 38.44%, respectively, while the contents of CaO and MgO in the coal ash are only 1% and 0.59%, respectively. The raw material for preparing residual char No. 2 is 20% palm kernel coke and 80% coal powder. Therefore, the content of SiO_2 and Al_2O_3 in residual char No. 2 will be lower than that of residual char No. 1, but the content of CaO and MgO will increase slightly. CaO and MgO are alkaline oxides, which can ionize a large amount of free oxygen ions (O^{2-}) into the slag, increasing the activity of O^{2-} in the slag, reducing the aggregation degree of Al-O anion groups, breaking their network structure, and forming simple mono- and di-tetrahedral structures. At the same time, the increase of O^{2-} can cause the decomposition of complex silicon-oxygen compound anions $Si_xO_y^{2-}$ in the slag, thus reducing the viscosity of the slag[35]. Zhang et al.[36] proposed that oxygen in slag exists in three forms: bridging oxygen (O^0), non-bridging oxygen (O^{-1}), and free oxygen (O^{-2}). Free oxygen (O^{-2}) exists in "MO", bridging oxygen (O^0) exists in SiO_2, while non-bridging oxygen (O^{-1}) is present in $2MO \cdot SiO_2$. When Al_2O_3 is present in the slag, M ions in MO will provide charge compensation for Al^{3+}, forming aluminum-oxygen tetrahedral units where oxygen exists in O^0 form. Furthermore, these tetrahedral units may enter into silicon-oxygen tetrahedral units, resulting in the formation of more complex alumino-silicate network structures in slag. Thus, an increase in the Al_2O_3 content in slag will lead to an increase in the complexity of the alumino-silicate network structure, thereby increasing the slag viscosity.

4.5 Study on the changes in smelting parameters of mixed injection of biomass char and pulverized coal into blast furnace

In order to analyze the degree of energy utilization in blast furnace smelting, people usually use calculation methods to determine the actual fuel ratio, coke ratio, gas utilization rate, etc. Furthermore, in order to delve deeper into the research, material balance and heat balance calculations, direct reduction degree calculations, theoretical coke ratio calculations, and the influence of various factors on coke ratio are used to discover issues and seek ways to further improve energy utilization. With the adoption of new technological measures in blast furnaces, such as high wind temperature, enriched oxygen, and various fuel injection methods, precise calculations can be used to predict smelting efficiency, and thus the most suitable smelting system can be formulated. For new blast furnace processes, such calculations are the important basis for the design and selection of the blast furnace body and its ancillary equipment, as well as the main basis for transportation and power balance in steel enterprises[37]. When palm shell semi-coke and coal powder are mixed and injected into the blast furnace, it can cause a series of effects on the smelting process, thus it is necessary to study the effects of the mixed injection on blast furnace smelting parameters, in order to provide a basis for the design of new blast furnace smelting processes.

4.5.1 Establishment of material and heat balance model for blast furnace smelting

4.5.1.1 Mass balance calculation

A Material consumption calculation

In order to determine the consumption of various materials during the blast furnace smelting process, typically Fe, P, Mn, V, Nb and other equilibrium equations are established based on the balance of required elements in pig iron composition[37]. Alkalinity balance equations are established based on the slag alkalinity and the designated content of slag oxidants, and are then solved simultaneously through joint calculation.

Fe balance:

$$m_{Sin} \cdot w(Fe_{Sin}) + m_{Pel} \cdot w(Fe_{Pel}) + m_{Sol} \cdot w(Fe_{Sol}) + m_{Coke} \cdot w(Fe_{Coke}) + m_{Coal} \cdot w(Fe_{Coal}) + m_{Bio} \cdot w(Fe_{Bio}) - m_{Dust} \cdot w(Fe_{Dust}) = 1000[Fe]/\eta_{Fe} \quad (4\text{-}17)$$

P balance:

$$m_{Sin} \cdot w(P_{Sin}) + m_{Pel} \cdot w(P_{Pel}) + m_{Sol} \cdot w(P_{Sol}) + m_{Coke} \cdot w(P_{Coke}) + m_{Coal} \cdot w(P_{Coal}) + m_{Bio} \cdot w(P_{Bio}) - m_{Dust} \cdot w(P_{Dust}) = 1000[P]/\eta_{P} \quad (4\text{-}18)$$

Alkalinity balance:

$$[m_{Sin} \cdot w(CaO_{Sin}) + m_{Pel} \cdot w(CaO_{Pel}) + m_{Sol} \cdot w(CaO_{Sol}) + m_{Coke} \cdot w(CaO_{Coke}) + m_{Coal} \cdot w(CaO_{Coal}) + m_{Bio} \cdot w(CaO_{Bio}) - m_{Dust} \cdot w(CaO_{Dust})]/$$
$$[m_{Sin} \cdot w(SiO_{2\text{-}Sin}) + m_{Pel} \cdot w(SiO_{2\text{-}Pel}) + m_{Sol} \cdot w(SiO_{2\text{-}Sol}) + m_{Coke} \cdot w(SiO_{2\text{-}Coke}) + m_{Coal} \cdot w(SiO_{2\text{-}Coal}) + m_{Bio} \cdot w(SiO_{2\text{-}Bio}) - m_{Dust} \cdot w(SiO_{2\text{-}Dust}) - 2.14 \times [Si] \times 10] = R$$
$$(4\text{-}19)$$

where, m_{Sin}, m_{Pel}, m_{Sol}, m_{Dust}, m_{Coke}, m_{Coal} and m_{Bio} are the amount of sinter, pellet, flux, furnace dust, coke, pulverized coal injection and biomass semi-coke, respectively, kg/tHM; $w(Fe_{Sin})$, $w(Fe_{Pel})$, $w(Fe_{Sol})$, $w(Fe_{Dust})$, $w(Fe_{Coke})$, $w(Fe_{Coal})$ and $w(Fe_{Bio})$ are iron content in sinter, pellets, flux, furnace dust, coke, pulverized coal and biomass semi-coke, respectively, %; $w(P_{Sin})$, $w(P_{Pel})$, $w(P_{Sol})$, $w(P_{Dust})$, $w(P_{Coke})$, $w(P_{Coal})$ and $w(P_{Bio})$ are P content in sinter, pellets, flux, furnace dust, coke, pulverized coal and biomass semi-coke, respectively, %; $w(CaO_{Sin})$, $w(CaO_{Pel})$, $w(CaO_{Sol})$, $w(CaO_{Dust})$, $w(CaO_{Coke})$, $w(CaO_{Coal})$ and $w(CaO_{Bio})$ are CaO content in sinter, pellets, flux, furnace dust, coke, pulverized coal and biomass semi-coke, respectively, %; $w(SiO_{2\text{-}Sin})$, $w(SiO_{2\text{-}Pel})$, $w(SiO_{2\text{-}Sol})$, $w(SiO_{2\text{-}Dust})$, $w(SiO_{2\text{-}Coke})$, $w(SiO_{2\text{-}Coal})$ and $w(SiO_{2\text{-}Bio})$ are SiO_2 content in sinter, pellets, flux, furnace dust, coke, pulverized coal and biomass semi-coke, respectively, %; [Fe], [P], [Si] are the content of Fe, P and Si in pig iron, respectively, %; η_{Fe}, η_P are the distribution ratio of Fe and P in slag and hot metal; R is slag alkalinity.

B Calculation of slag amount and slag composition

According to the distribution of elements between slag iron, the reduced oxide and sulfide formed by desulfurization are calculated, and the total amount is the amount of slag[37]:

$$w(SiO_2) = \sum m_i \cdot w(SiO_{2\text{-}i}) - 2.14 \times [Si] \times 10 \qquad (4\text{-}20)$$

$$w(CaO) = \sum m_i \cdot w(CaO_i) \qquad (4\text{-}21)$$

$$w(MgO) = \sum m_i \cdot w(MgO_i) \qquad (4\text{-}22)$$

$$w(Al_2O_3) = \sum m_i \cdot w(Al_2O_{3\text{-}i}) \qquad (4\text{-}23)$$

$$w(MnO) = [Mn] \times 10 \times \frac{71}{55} \times \frac{1-\eta_{Mn}}{\eta_{Mn}} \qquad (4\text{-}24)$$

$$w(FeO) = [Fe] \times 10 \times \frac{72}{56} \times \frac{1-\eta_{Fe}}{\eta_{Fe}} \qquad (4\text{-}25)$$

$$\frac{1}{2}w(S) = \{w(S_{Burden}) \times 0.95 - [S] \times 10\}/2 \qquad (4\text{-}26)$$

$$m_{Slag} = w(SiO_2) + w(CaO) + w(MgO) + w(Al_2O_3) + w(MnO) + w(FeO) + \frac{1}{2}w(S) \qquad (4\text{-}27)$$

where, m_i is charge consumption, kg/tHM; $w(SiO_2)$, $w(CaO)$, $w(MgO)$, $w(Al_2O_3)$, $w(MnO)$, $w(FeO)$ are the content of each oxide of the slag, respectively, kg/tHM; η_{Mn} is the partition coefficient of Mn in slag iron; $w(S_{Burden})$ is the total amount of sulfur brought in by the charge, kg/tHM; [S] is the amount of sulfur in molten iron, %; m_{Slag} is the amount of slag generated, kg/tHM. The amount of each oxide is divided by the total amount of slag to obtain its content in the slag or slag composition.

C Blast volume calculation

The amount of carbon burned at the air outlet is calculated by solving for the carbon balance.

$$m_{C\text{-}R} = m_{C\text{-}Coke} + m_{C\text{-}Coal} + m_{C\text{-}Bio} - m_{C\text{-}dFe} - m_{C\text{-}Si,Mn,P,S} - m_{C\text{-}pig} \qquad (4\text{-}28)$$

The concentration of oxygen in the blast is:

$$\beta_{O_2\text{-}B} = 0.21 \times (1 - \phi_{H_2O}) + 0.5\phi_{H_2O} \qquad (4\text{-}29)$$

The amount of blast air required for C combustion before the air outlet is:

$$V_B = [m_{C\text{-}R} \times 22.4/(2 \times 12) - m_{Coal}\beta_{O\text{-}Coal} - m_{Bio}\beta_{O\text{-}Bio} \times (22.4/16)]/\beta_{O_2\text{-}B} \qquad (4\text{-}30)$$

where, $m_{C\text{-}R}$ is the mass of carbon burned in front of the air vent, kg/tHM; $m_{C\text{-}Coke}$, $m_{C\text{-}Coal}$, $m_{C\text{-}Bio}$, $m_{C\text{-}dFe}$, $m_{C\text{-}Si,Mn,P,S}$, $m_{C\text{-}pig}$ are the carbon content of coke, pulverized coal and biomass semi-coke, iron directly reduces the carbon consumed, Si, Mn, P, S directly reduce the amount of carburizing of carbon and molten iron consumed, respectively, kg/tHM; $\beta_{O_2\text{-}B}$ is the oxygen content of the blast, %; $\beta_{O\text{-}Coal}$ is pulverized coal oxygen content, %; $\beta_{O\text{-}Bio}$ is biomass semi-coke oxygen content,%.

D The amount of gas on the roof and the composition of the gas

The main components of furnace roof gas are composed of CO_2, H_2, CO, H_2O and N_2, and the volume of each component is calculated by the following formula:

$$V_{CO_2}^T = V_{CO_2}^{Fe_2O_3} + V_{CO_2}^{Fe} + V_{CO_2}^{Vol} \quad (4\text{-}31)$$

$$V_{H_2}^T = \left[V_B \cdot \phi_{H_2O} + \left(\frac{m_{Coal}\beta_{H\text{-}Coal} + m_{Bio}\beta_{H\text{-}Bio} + m_{Coke}\beta_{H\text{-}Coke}}{2} + \sum m_i \cdot \beta_{i\text{-}H_2O} \right) \times 22.4 \right] \times (1 - \eta_{H_2})$$

$$(4\text{-}32)$$

$$V_{CO}^T = V_{CO}^R + (m_{C\text{-}dFe} + m_{C\text{-}Si,Mn,P,S}) \times 22.4 - V_{CO_2}^{Fe_2O_3} - V_{CO_2}^{Fe} \quad (4\text{-}33)$$

$$V_{H_2O}^T = \left[V_B \cdot \phi_{H_2O} + \left(\frac{m_{Coal}\beta_{H\text{-}Coal} + m_{Bio}\beta_{H\text{-}Bio} + m_{Coke}\beta_{H\text{-}Coke}}{2} + \sum m_i \cdot \beta_{i\text{-}H_2O} \right) \times 22.4 \right] \times \eta_{H_2}$$

$$(4\text{-}34)$$

$$V_{N_2}^T = V_B \cdot (1 - \beta_{O_2\text{-}B}) \quad (4\text{-}35)$$

$$V_{Gas}^{Top} = V_{CO_2}^T + V_{H_2}^T + V_{CO}^T + V_{H_2O}^T + V_{N_2}^T \quad (4\text{-}36)$$

where, V_{Gas}^{Top}, $V_{CO_2}^T$, $V_{H_2}^T$, V_{CO}^T, $V_{H_2O}^T$, $V_{N_2}^T$ are the volume of the top gas and the volume of each component, m³/tHM; $V_{CO_2}^{Fe_2O_3}$ is high-valent iron, manganese oxides, etc. are reduced to low-valent oxides to form CO_2, m³/tHM; $V_{CO_2}^{Fe}$ is indirect reduction of CO_2 generated by Fe, m³/tHM; $V_{CO_2}^{Vol}$ is CO_2 generated by coke volatilization, m³/tHM; $\beta_{H\text{-}Coal}$, $\beta_{H\text{-}Bio}$, $\beta_{H\text{-}Coke}$ are hydrogen content of pulverized coal injection, biomass semi-coke and coke,%; $\beta_{i\text{-}H_2O}$ is the content of crystalline water for sinter, pellets and lumps,%.

E Material balance sheet preparation

The consumption of raw materials and blast air by a unit of smelted pig iron are regarded as income items in the material balance, while the smelted pig iron, along with the calculated amounts of slag, top gas, and furnace dust, are considered as expense items. Under scientifically sound calculation methods and with accurate raw data, the income and expense items in the material balance are equal. However, due to various errors, actual calculations may deviate from the theoretical balance. Current testing and statistical methods require that the relative error between income and expense items shall not exceed 0.3%.

4.5.1.2 Heat balance calculation

Thermal balance calculation of the pig iron production process in the blast furnace involves taking into account the heat input, which comes from the combustion of carbon in the tuyere combustion zone generating heat through the release of CO, and the physical heat brought into the interior of the blast furnace by the hot air. The corresponding heat output items include direct reduction heat consumption and others. The advantage of this method is that it clearly shows the effect of direct reduction on heat consumption in the heat balance, and this part of the heat consumption is mainly compensated by the heat released from the combustion of carbon to CO in the tuyere. Therefore, this method can demonstrate the effect of direct reduction on the fuel ratio[29].

A Hot revenue item calculation

Carbon combustion and heat release before air vent:

$$Q_C^R = m_{C\text{-}R} \cdot q_{C\text{-}CO} \tag{4-37}$$

where $q_{C\text{-}CO}$ ——Oxidation of carbon per unit mass to generate CO releases heat, kJ/kg.

Hot air brings physical heat:

$$Q_B^R = \int_{T_0}^{T} \frac{V_B \cdot c_B}{22.4} dT - q_{H_2O\text{-}H_2} \cdot V_B \cdot \phi_{H_2O} \tag{4-38}$$

where c_B ——Blow heat capacity, J/(K · mol);

T ——Hot air temperature, K;

$q_{H_2O\text{-}H_2}$ ——Heat dissipation of water decomposition, kJ/m³.

B Heat expense item calculation

Direct reduction of heat consumption:

$$Q_d = m_{C\text{-}dFe} \cdot q_{C\text{-}dFe} + \sum m_{C\text{-}i} \cdot q_{C\text{-}i} \tag{4-39}$$

where $q_{C\text{-}dFe}$ ——Reduces the heat consumed per unit mass of iron, kJ/kg;

q_{C-i} ——Non-ferrous elements per unit mass are directly reduced to consume calories, kJ/kg, where i are Si, Mn, P and S.

Heat consumption for desulfurization:

$$Q_S = q_S \cdot m_S \tag{4-40}$$

where m_S ——Desulfurization amount, kg/tHM;

q_S ——Removes the heat consumed by sulfur per unit mass, kJ/kg.

Slag molten iron takes away heat:

$$Q_{Slag} = m_{Slag} \cdot q_{Slag} \tag{4-41}$$

$$Q_{Pig} = 1000 q_{Pig} \tag{4-42}$$

where q_{Slag}, q_{Pig} ——Specific heat capacity of slag and pig iron, kJ/kg.

Heat taken away by the top gas:

$$Q_{Gas}^{Top} = \int_{T_0}^{T_{Top}} \frac{V_i \cdot c_i}{22.4} dT \tag{4-43}$$

where V_i, c_i ——The composition of the roof gas and the heat capacity, m³/tHM, J/(K · mol);

T_{Top} ——The temperature of the top gas, K.

In the calculation, the value of the heat input item of the blast furnace process should be greater than that of the heat output item. The difference between the heat input item and the heat output item resulting in the thermal value represents the heat loss during the blast furnace smelting process, which is mainly caused by cooling water and heat dissipation from the furnace body.

4.5.1.3 Determination of smelting conditions

A Composition of raw materials

The raw material data used in this section's calculation is provided by a certain domestic steel enterprise, and the composition of raw materials is listed in Table 4-19 and Table 4-20, while the composition of furnace dust is shown in Table 4-21.

Table 4-19 The main chemical composition of the raw materials entering the furnace (%)

Raw materials	TFe	FeO	CaO	SiO$_2$	MgO	Al$_2$O$_3$	MnO$_2$	MnO	S	P
Sinter	54.95	8.63	10.27	5.86	2.82	2.19	—	0.11	0.02	0.04
Pellets	62.93	3.76	0.78	7.35	0.54	1.20	—	0.27	0.02	0.09

Table 4-20 Fuel composition in the furnace (%)

Composition	C$_F$	S	A						V	M
			CaO	SiO$_2$	MgO	Al$_2$O$_3$	Other	Total		
Coke	86.60	0.85	0.52	5.75	0.17	4.38	1.58	12.40	0.98	0.02
Coal	79.94	0.67	0.43	6.11	0.07	4.60	0.6	11.81	7.48	0.84
Semi-focal	79.82	0.08	2.80	0.21	4.16	0.07	3.59	10.83	7.51	1.84

Table 4-21 Furnace dust composition (%)

Furnace dust	TFe	FeO	CaO	SiO$_2$	MgO	Al$_2$O$_3$	MnO$_2$	MnO	S	P
Value	36.94	5.20	7.34	4.30	1.46	2.61	—	0.12	0.34	0.03

B The preset composition of pig iron

The preset composition of pig iron is shown in Table 4-22, and the parameters of the smelting process are set accordingly $T_{Pig} = 1,450℃$. The distribution ratio of each element in slag and iron is listed in Table 4-23.

Table 4-22 Preset pig iron composition (%)

Element	Fe	C	Si	Mn	P	S
Content	94.825	4.445	0.470	0.150	0.086	0.024

Table 4-23 Allocation ratio of each element

Products	Fe	Mn	P	S
Pig iron	0.9971	0.9	1	0.1
Slag	0.0029	0.1	—	0.9

4.5.1.4 Data collation and calculation

Based on the raw material conditions provided by the enterprise, the blast furnace fuel ratio, coal ratio, oxygen enrichment rate, blast humidity, blast capacity, wind temperature, slag alkalinity listed in Table 4-24 are taken as input variables, while the blast furnace gas utilization rate, utilization coefficient, material balance, and heat balance relationships are taken as verification conditions to solve the r_d^0 value.

Table 4-24 Underlying data

Coke ratio (kg/tHM)	Coal ratio (kg/tHM)	Oxygen enrichment (%)	Blast humidity (%)	Blow ability (m³/min)
380	140	1.0	1.0	9,000
Blast temperature (℃)	Furnace gas temperature (℃)	Slag alkalinity	Slag volume (kg/tHM)	Gas utilization (%)
1,150	159	1.15	360	46.0

In the material balance and heat balance calculations, a complete chemical composition of the raw materials is required. The composition of sintered ore and pellet ore provided in the previous chapter only includes the contents of TFe, P, and S elements, as well as several compound components such as FeO, CaO, SiO_2, MgO, and Al_2O_3. In actual calculations, it is necessary to first preliminarily organize the chemical composition of the raw materials, determine the state of each element, and establish the actual percentage content of each component.

It is generally believed that Fe exists in the form of Fe_2O_3 and FeO in sintered ore and pellet ore. The Fe_2O_3 content in the ore can be calculated using the following formula under the condition of determining the TFe content and FeO content:

$$w(Fe_2O_3) = \left([TFe] - \frac{56}{72} \times w(FeO) \right) \times \frac{160}{112} \qquad (4-44)$$

Phosphorus exists in the form of P_2O_5 in ores, and the calculation formula for P_2O_5 content is as follows:

$$w(P_2O_5) = [P] \times \frac{142}{62} \qquad (4-45)$$

The S content in ores is calculated as the S/2 content. In sintered ore, S exists in the form of FeS, and Fe has already been calculated in the FeO item. Since the atomic weight of oxygen is only half that of sulfur, half of the remaining S content in the compositional arrangement should be represented as S/2.

After sorting, the composition of sintered ore and pellet ore is shown in Table 4-25.

Table 4-25 Finished raw material composition   (%)

Raw material	S/2	P_2O_5	Fe_2O_3	FeO	CaO	SiO_2	MgO	Al_2O_3	MnO	MnO_2
Sinter	0.01	0.10	68.91	8.63	10.27	5.86	2.82	2.19	0.11	—
Pellets	0.01	0.20	85.72	3.76	0.78	7.35	0.54	1.2	0.27	—

By utilizing organized raw material components, the calculation of blast furnace material balance and heat balance can be achieved. The revenue items in material balance mainly consist of coke, injected fuel, ore, and blast volume, while expenditure items primarily include pig iron, slag, gas, and furnace dust. A material balance sheet is compiled as shown in Table 4-26. On the basis of material balance, a heat balance is prepared as presented in Table 4-27. The composition of furnace top gas is illustrated in Table 4-28.

Table 4-26 Material balance sheet

Income	Amount (kg/tHM)	Expenditure item	Amount (kg/tHM)
Ore	1,683.9	Pig iron	1,000
Coke	380	Slag	367.3
Coal	140	Gas	2,313.5
Blast volume	1,496.6	Dust	18
Total	3,700.5	Total	3,698.8
Absolute error (%)	1.7	Relative error (%)	0.05

Table 4-27 Heat balance table

Income items	Heat (MJ/tHM)	Percentage (%)	Expense items	Heat (MJ/tHM)	Percentage (%)
Burn carbon in front of the air outlet to release heat	2,572.8	59.51	Direct restore heat consumption	1,457.4	33.71
			Heat consumption for desulfurization	19.5	0.45
Hot air brings physical heat	1,750.7	40.49	Carbonate decomposition consumes heat	0	0
			The slag takes away the heat	653.7	15.12
			Molten iron takes away heat	1,240	28.68
			Top gas takes away heat	340.9	7.88
			Heat loss	612	14.16
Total	4,323.5	100	Total	4,323.5	100

Table 4-28 Top gas composition

CO(%)	H_2(%)	N_2(%)	CO_2(%)	H_2O(%)	Σ(%)	V(m³)
23.18	2.02	52.99	19.79	2.02	100	1,702.50

Based on Tables 4-27 and 4-28, the calculated slag weight is 367.3kg/tHM, the gas utilization rate is 46.1%, and the top gas temperature is 165℃. The similarity of the calculated slag weight and gas utilization rate to the basic data provided by the enterprise confirms the accuracy of the calculation process. Under the original raw material and operating parameter conditions in this article, A is determined to be 0.51. The benchmark blast furnace smelting parameters for the reference period are shown in Table 4-29.

4.5 Study on the changes in smelting parameters of mixed injection of biomass char and pulverized coal into blast furnace

Table 4-29 Basic parameters of blast furnace smelting in reference period

Parameter	Unit	Symbol	Value
Theoretical combustion temperature	℃	T_f	2,246
Slag volume	kg/tHM	m_{Slag}	367.3
Gas utilization	%	η_{CO}	46.1
Furnace top coal temperature	℃	T_{Top}	165
Utilization factor	$t/(m^3 \cdot d)$	λ	3.40

4.5.2 Changes in smelting parameters of blast furnace mixed injection biomass semi-coke and pulverized coal

In order to explore the influence of mixed injection palm shell semi-coke and pulverized coal on blast furnace smelting, the established material equilibrium and thermal equilibrium models were used to calculate the changes of blast furnace smelting parameters after adding palm shell semi-coke of different masses. The composition of palm shell semi-coke is listed in Table 4-20.

4.5.2.1 Direct reduction degree and fuel ratio

There is no good way to design the direct reduction degree of the blast furnace, currently. This section uses the empirical formula proposed by Professor A. H. Ram [38] to calculate the change of direct reduction of blast furnace with the amount of semi-coke injection of palm shell. After the blast furnace sprayed palm shell semi-coke, the reducing components in the gas increased, and H_2 in terms of reducing thermodynamics and kinetics, was better than CO, therefore, after the blast furnace sprayed palm bright semi-coke, indirect reduction in the furnace was developed, resulting in a decrease in direct reduction.

Because palm shell semi-coke is considered to replace part of the pulverized coal for injection, the coke ratio is fixed in the calculation of this section, and the influence of injection palm shell semi-coke on the fuel ratio is investigated under the condition that the heat loss of tons of iron remains unchanged. After blast furnace injection palm shell semi-coke, the change of fuel ratio is shown in Fig. 4-37, with the increase of palm shell semi-coke injection, the fuel ratio shows a downward trend, mainly because after blast furnace injection palm shell semi-coke, the direct reduction degree decreases, resulting in the reduction of carbon consumption by direct reduction, and the mixing of palm shell semi-coke and pulverized coal can promote pulverized coal combustion burning, improve the combustion of pulverized coal in the raceway, increase the combustion rate of pulverized coal, thereby reducing the fuel ratio of the blast furnace.

4.5.2.2 Amount of air outlet coke and gas utilization rate

The main role of coke in the blast furnace can be divided into four aspects: (1) reducing agent;

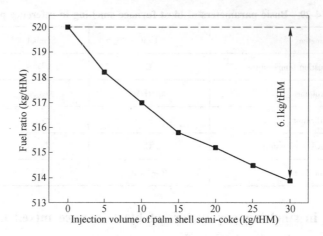

Fig. 4-37 Effect of blast furnace injection palm shell semi-coke on fuel ratio

(2) heating agent; (3) carburizing agent; (4) skeleton effect. The quality of coke burned at the tuyere is an important parameter to judge the role of coke as a skeleton inside the blast furnace, the quality of coke burned before the tuyere is reduced, and the amount of coke acting as a skeleton role is increased, which can ensure the air permeability of blast furnace smelting and play an important role in the stability and downstream of the blast furnace. Fig. 4-38 shows the influence of blast furnace injection palm shell semi-coke on the number of tuyere coke, and the blast furnace injection palm shell semi-coke is promoted indirect reduction in the furnace, so that the degree of direct reduction is reduced, resulting in a decrease in the quality of coke participating in direct reduction, and an increase in the amount of coke that can be burned before reaching the tuyere. When the semi-coke spray volume of palm shell reaches 30kg/tHM, the quality of tuyere coke increases by 2.47kg/tHM, which will not adversely affect the blast furnace.

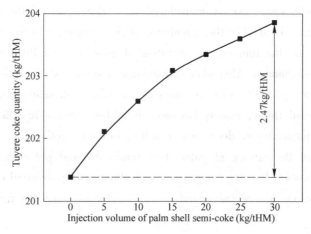

Fig. 4-38 Effect of blast furnace injection palm shell semi-coke on the number of air outlet coke

The formula for gas utilization is:

$$\eta_{CO} = \frac{\varphi_{CO_2}}{\varphi_{CO_2} + \varphi_{CO}} \qquad (4\text{-}46)$$

where φ_{CO_2}——Volume fraction of CO_2 in top gas, %;

φ_{CO}——Volume fraction of CO in top gas, %.

After the blast furnace sprayed palm shell semi-coke, the direct finger reduction degree decreased, and the indirect reduction of the upper area of the blast furnace was developed, resulting in a decrease in the CO ratio in the furnace roof gas and an increase in the CO_2 ratio, so the gas utilization rate increased after the blast furnace sprayed palm shell semi-coke, as shown in Fig. 4-39.

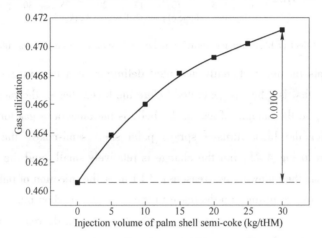

Fig. 4-39 Effect of blast furnace injection palm shell semi-coke on gas utilization rate

4.5.2.3 Tons of iron blast air and belly gas volume

Fig. 4-40 shows the effect of blast furnace injection palm shell semi-coke on ton iron blast volume, and the influence of injection palm shell semi-coke on ton iron blast volume mainly has two aspects: (1) after blast furnace injection palm shell semi-coke, the coal ratio decreases, and the coal ratio decreases more than the increase of injection palm shell semi-coke, resulting in a decrease in fuel ratio, so that the amount of carbon burned before the tuyere is reduced, and the amount of oxygen required for combustion is reduced, resulting in a decrease in the ton iron blast volume; (2) blast furnace injection palm shell semi-coke reduces the direct reduction, the reduction of direct reduction reduces the amount of coke consumed by direct reduction, and the quality of coke burned before reaching the air outlet increases, so that the air volume of tons of iron has a tendency to increase. However, the decrease in carbon burned in the raceway caused by the decrease in the coal ratio is greater than the increase in the amount of tuyere coke, and the amount of oxygen required for carbon combustion at the tuyere is reduced, so it will eventually lead to a decrease in the air volume of tons of iron. For every 10kg increase in the semi-coke blowing volume of palm shell, the air volume of ton iron blast decreases by 1.2-3.5m³/tHM, and the decrease decreases with the increase of palm shell semi-coke spray volume.

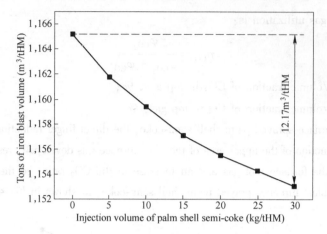

Fig. 4-40　Effect of blast furnace spraying palm shell semi-coke on ton iron blast volume

This section adopts the internationally accepted definition of a certain amount of coal in the belly of the furnace, that is, the gas generated before the tuyere leaves the raceway and enters the belly of the furnace, so the amount of gas in the belly is the amount of gas formed by combustion in the raceway. After the blast furnace sprays palm shell semi-coke, the belly gas volume decreases, as shown in Fig. 4-41, but the change is relatively small, and the change of the belly gas volume mainly has the following two reasons: (1) after the injection of palm shell semi-coke, the fuel ratio decreases, resulting in a decrease in the amount of carbon burned before the tuyere, resulting in a decrease in the CO produced by combustion, and a decrease in the gas volume in the belly of the furnace; (2) after the injection of palm shell semi-coke, due to the reduction of the air volume of tons of iron and the decrease in N_2 carried by the blaster, it will also lead to a decrease in the gas volume of the belly.

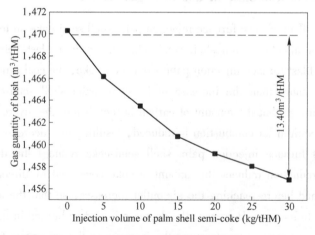

Fig. 4-41　Effect of blast furnace injection palm shell semi-coke on gas volume in furnace belly

4.5.2.4　Theoretical combustion temperature

The theoretical combustion temperature is an important parameter to characterize the thermal state of the furnace cylinder, the theoretical combustion temperature is too low, the heat of the furnace cylinder is insufficient, and the melting and dripping temperature of slag iron is not enough, which will cause the furnace cylinder to accumulate. The increase of theoretical combustion temperature and the concentration of heat in the furnace cylinder are conducive to the progress of the smelting reaction, but the theoretical combustion temperature is too high, which will increase the heat load of the furnace cylinder, affect the air permeability of the charge, and is not conducive to the stable production of the blast furnace. After the domestic blast furnace injection fuel, it is considered that the theoretical combustion temperature is more reasonable to maintain between 2,000-2,300℃ [37]. It can be seen from Fig. 4-42 that after the blast furnace injection palm shell semi-coke, the theoretical combustion temperature changes very little and basically remains stable, and the influence of the palm shell semi-coke on the theoretical combustion temperature mainly has the following three aspects: (1) the air volume of ton iron blast is reduced, and the physical heat brought in by the hot air is reduced; (2) the fuel ratio is reduced, and the heat consumption of pulverized coal decomposition is reduced; (3) the amount of gas in the belly of the furnace is reduced, and the heat consumed by heating gas is reduced. Under the comprehensive influence of several aspects, the theoretical combustion temperature remained basically stable.

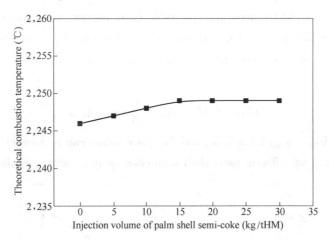

Fig. 4-42　Effect of blast furnace injection palm shell semi-joke on theoretical combustion temperature

4.5.2.5　Calculation of CO_2 emission reduction

For blast furnace ironmaking processes, the fuel ratio determines the final carbon footprint. From the above mass balance calculation, it can be concluded that the blast furnace injection palm shell semi-coke can reduce the coal and fuel ratio with the fixed focal ratio unchanged. In order to quantitatively analyze the potential of CO_2 reduction after the application of palm shell semi-coke in blast furnace injection process, the CO_2 emissions of palm shell semi-coke after injection were calculated.

There is a following relationship between the CO_2 emissions E_{CO_2} (kg/tHM) of the blast furnace and the fuel consumption M_f(kg/tHM):

$$E_{CO_2} = M_f \tau \qquad (4\text{-}47)$$

where, τ is the CO_2 emission coefficient of solid carbon fuel, that is, the CO_2 emissions (kg/kg) produced by the complete combustion of carbon fuel per unit mass, for pure carbon, the value of τ in the case of complete combustion is $\tau = \dfrac{M_{CO_2}}{M_C} = \dfrac{44}{12} = \dfrac{11}{3}$ kg/kg, but in fact, the carbon content of carbon fuel and the ratio of CO_2 generated by complete combustion are less than 100%, for fuel with carbon content w (%), its combustion rate is η (%), then the CO_2 emission E_{CO_2} of the blast furnace is calculated according to Eq. (4-48):

$$E_{CO_2} = M_f \tau w \eta = \frac{11}{3} M_f w \eta \qquad (4\text{-}48)$$

Suppose that when the blast furnace injection is carried out with palm shell semi-coke, the amount of pulverized coal for Tish is x (kg/tHM), then bring CO_2 emission reductions can be calculated according to Eq. (4-49):

$$\Delta E_{CO_2} = \Delta E_{CO_2\text{-Coal}} - E_{CO_2\text{-Bio}} = \frac{11}{3} x w_{Coal} \eta_{Coal} - \frac{11}{3} M_{f\text{-Bio}} w_{Bio} \eta_{Bio} \qquad (4\text{-}49)$$

Biomass is considered to be green and renewable clean energy, absorbing CO_2 during growth, which can offset the CO_2 produced during combustion consumption, assuming $E_{CO_2\text{-Bio}} = 0$, then the CO_2 emission reduction brought by blast furnace injection palm shell semi-coke is calculated as:

$$\Delta E_{CO_2} = \Delta E_{CO_2\text{-Coal}} = \frac{11}{3} x w_{Coal} \eta_{Coal} \qquad (4\text{-}50)$$

where, w_{Coal} is 79.94%; η_{Coal} takes 80%, and the above values can be brought into the Eq. (4-47) to calculate the ΔE_{CO_2} of different palm shell semi-coke spray amounts, as shown in Fig. 4-43.

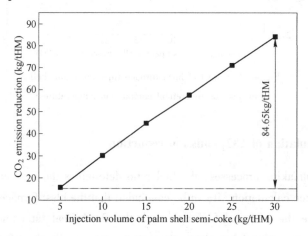

Fig. 4-43 Relationship between palm shell semi-focal injection and CO_2 emission reduction

With the increase of palm shell semi-focal spray amount, CO_2 emission reduction gradually increased, when the palm shell semi-coke injection amount is 30kg/tHM, CO_2 84.65kg/tHM can be reduced. Therefore, the replacement of some pulverized coal for blast furnace injection in palm shell semi-coke can effectively reduce the dependence of blast furnace ironmaking on fossil energy, reduce coal mining, effectively reduce CO_2 and emissions of blast furnace, alleviate the dilemma of ironmaking process, and is of great significance for achieving the national CO_2 emission reduction target.

4.6 Chapter summary

(1) With the increase of pyrolysis temperature, the yield of biomass semi-coke gradually increases. The composition and structure of different biomass semi-coke depend on the pyrolysis process parameters and the composition of the initial biomass. N_2 adsorption technology was used to characterize the pore structure, and it was found that the biomass semi-coke particles had a continuous and complete pore system, as small as the molecular level, as large as no upper limit pores, and the proportion of micropores and mesopores was large. Analyzing the microscopic morphology of semi-focal particles, it is found that the palm shell semi-focal structure is mostly irregular geometric block structure, with the increase of pyrolysis temperature, its own pore structure first develops and then melts and collapses, and the soybean straw semi-coke structure is mostly hollow tubular structure, with the increase of pyrolysis temperature, the pore structure first remains unchanged and then gradually develops. The semi-coke particles of corn cob exhibit a layered skeleton structure under the microscopic level, and the pore structure gradually melts and disappears with the increase of pyrolysis temperature.

(2) The combustion reactivity of palm shell semi-coke is better than that of pulverized coal, mainly because palm shell semi-coke has a pore structure that is more developed than pulverized coal, the content of alkaline substances with catalytic effect in ash is greater than that of pulverized coal, and the order degree and graphitization degree of semi-coke carbon structure of palm shell are small and dry coal powder. After the mixed combustion of palm shell semi-coke and pulverized coal, the combustion characteristics of pulverized coal were improved, and there was a synergistic effect in the mixed combustion process, the presence of pulverized coal limited the ignition and combustion of palm shell semi-coke, but the heat released after palm shell semi-coke combustion promoted the fire of pulverized coal, and the alkali metal compounds in the semi-coke ash of palm shell in the later stage promoted the combustion of carbon in pulverized coal. The activation energy range of the mixed combustion reaction of palm shell semi-coke and pulverized coal calculated by the random pore model was 90.2-121.8kJ/mol. In order to control the alkali load in the furnace, the proportion of palm shell semi-coke should be less than 20%.

(3) The unburned residual carbon produced after the blast furnace sprays the palm shell semi-coke is more reactive than that of coke, which can protect the coke. Traditional unburned pulverized coal and unburned biomass residual carbon exist in the slag in the form of solid particles, which will increase the viscosity of the slag. However, the contents of CaO and MgO in

the semi-coke ash of palm shell are greater than those in pulverized coal, and compared with pulverized coal residual carbon, the influence of palm shell semi-coke unburned residual carbon on the viscosity of slag is small.

(4) Through the calculation of the mass equilibrium and thermal equilibrium of the blast furnace, it is found that after the blast furnace sprays the palm shell semi-coke, the direct reduction degree will be reduced, the fuel ratio will be reduced, and when the palm shell semi-coke injection volume is 30kg/tHM, the fuel ratio will be reduced by 6.1kg/tHM. When the half-coke injection volume of palm shell reached 30kg/tHM, the coke quality of the tuyere increased by 2.47kg/tHM, and the gas utilization rate increased from 0.46 to 0.47. For every 10kg increase in the semi-coke blowing volume of palm shell, the air volume of ton iron blast decreases by 1.2-3.5m^3/tHM, and the decrease decreases with the increase of palm shell semi-coke injection volume, and the theoretical combustion temperature remains basically unchanged. With the increase of palm shell bright semi-focal spray volume, the CO_2 emission reduction gradually increases, and when the palm shell semi-coke injection amount is 30kg/tHM, CO 284.65kg/tHM can be reduced. In general, in theory, it is feasible to replace some pulverized coal for blast furnace injection, and blast furnace injection palm shell semi-coke is conducive to the development of indirect reduction in the furnace, reduce fuel consumption, and effectively reduce CO_2 emissions.

References

[1] Zhou Zhongren, Wu Wenliang. Status quo and prospects of biomass energy [J]. Transactions of the CSAE, 2005, 21 (12): 12-15.

[2] Xiao Jun, Duan Jingchun, Wang Hua, et al. The Present Situation of Using Biomass [J]. Safety and Environmental Engineering, 2003, 10 (1): 11-14.

[3] Bi Xuegong, Rao Changrun, Peng Wei. Development status of simultaneous Injection of Agricultural and forestry residues and its perspective [J]. Henan Metallurgy, 2012, 20 (3): 1-5.

[4] Xiong Wei, Wang Guoqiang, Zhou Shaoxuan. Comparison of energy consumption and environmental impact of replacement of coal with straw injection into blast furnace [J]. Environmental Science & Technology, 2013, 36 (4): 137-140.

[5] Babich A, Senk D, Fernandez M. Charcoal behaviour by its injection into the modern blast furnace [J]. ISIJ International, 2010, 50 (1): 81-88.

[6] Mathieson J G, Rogers H, Somerville M A, et al. Reducing net CO_2 emissions using charcoal as a blast furnace tuyere injectant [J]. ISIJ International, 2012, 52 (8): 1489-1496.

[7] de Castro J A, de Silva A J, Sasaki Y, et al. A Six-phases 3-D model to study simultaneous injection of high rates of pulverized coal and charcoal into the blast furnace with oxygen enrichment [J]. ISIJ International, 2011, 51 (5): 748-758.

[8] de Castro J A, de Mattos Araújo G, da Mota I O, et al. Analysis of the combined injection of pulverized coal and charcoal into large blast furnaces [J]. Journal of Materials Research and Technology, 2013, 2 (4): 308-314.

[9] Du S W, Chen W H, Lucas J A. Pretreatment of biomass by torrefaction and carbonization for coal blend used in pulverized coal injection [J]. Bioresource Technology, 2014, 161: 333-339.

[10] Wang C, Larsson M, Lövgren J, et al. Injection of solid biomass products into the blast furnace and its potential effects on an integrated steel plant [J]. Energy Procedia, 2014, 61: 2184-2187.

[11] Mundike L, Collard F X, Görgens J F. Co-combustion characteristics of coal with invasive alien plant chars prepared by torrefaction or slow pyrolysis [J]. Fuel, 2018, 225: 62-70.

[12] Wang G W, Zhang J L, Shao J G, et al. Thermal behavior and kinetic analysis of co-combustion of waste biomass/low rank coal blends [J]. Energy Conversion and Management, 2016, 124: 414-426.

[13] Duman G, Uddin M A, Yanik J. The effect of char properties on gasification reactivity [J]. Fuel Processing Technology, 2014, 118 (2): 75-81.

[14] Le Manquais K, Snape C, Barker J, et al. TGA and drop tube furnace investigation of alkali and alkaline earth metal compounds as coal combustion additives [J]. Energy and Fuels, 2012, 26 (3): 1531-1539.

[15] Ding L, Zhang Y, Wang Z, et al. Interaction and its induced inhibiting or synergistic effects during co-gasification of coal char and biomass char [J]. Bioresource Technology, 2014, 173: 11-20.

[16] Farrow T S, Sun C, Snape C E. Impact of biomass char on coal char burn-out under air and oxy-fuel conditions [J]. Fuel, 2013, 114 (6): 128-134.

[17] Edreis E M A, Luo G, Li A, et al. Synergistic effects and kinetics thermal behaviour of petroleum coke/biomass blends during H_2O co-gasification [J]. Energy Conversion and Management, 2014, 79: 355-366.

[18] Wang G W, Zhang J L, Zhang G H, et al. Experimental and kinetic studies on co-gasification of petroleum coke and biomass char blends [J]. Energy, 2017, 131: 27-40.

[19] Hu S, Ma X, Lin Y, et al. Thermogravimetric analysis of the co-combustion of paper mill sludge and municipal solid waste [J]. Energy Conversion and Management, 2015, 99: 112-118.

[20] Yildiz Z, Uzun H, Ceylan S, et al. Application of artificial neural networks to co-combustion of hazelnut husk-lignite coal blends [J]. Bioresource Technology, 2016, 200: 42-47.

[21] Huang Y W, Chen M Q, Luo H F. Nonisothermal torrefaction kinetics of sewage sludge using the simplified distributed activation energy model [J]. Chemical Engineering Journal, 2016, 298: 154-161.

[22] Fan C, Zan C, Zhang Q, et al. The oxidation of heavy oil: Thermogravimetric analysis and non-isothermal kinetics using the distributed activation energy model [J]. Fuel Process Technology, 2014, 119: 146-150.

[23] Fatehi H, Bai X S. Structural evolution of biomass char and its effect on the gasification rate [J]. Applied Energy, 2017, 185: 998-1006.

[24] Tuinstra F, Koenig J L. Raman spectrum of graphite [J]. The Journal of Chemical Physics, 1970, 53 (3): 1126-1130.

[25] Green P D, Johnson C A, Thomas K M. Applications of laser Raman microprobe spectroscopy to the characterization of coals and cokes [J]. Fuel, 1983, 62 (9): 1013-1023.

[26] Sadezky A, Muckenhuber H, Grothe H, et al. Raman microspectroscopy of soot and related car-bonaceous materials: spectral analysis and structural information [J]. Carbon, 2005, 43 (8): 1731-1742.

[27] Katagiri G, Ishida H, Ishitani A. Raman spectra of graphite edge planes [J]. Carbon, 1988, 26 (4): 565-571.

[28] Beyssac O, Goffé B, Petitet J P, et al. On the characterization of disordered and heterogeneous carbonaceous materials by Raman spectroscopy [J]. Spectrochimica Acta Part A: Molecular and Biomolecular Spectroscopy, 2003, 59 (10): 2267-2276.

[29] Cuesta A, Dhamelincourt P, Laureyns J, et al. Raman microprobe studies on carbon materials [J]. Carbon, 1994, 32 (8): 1523-1532.

[30] Bar-Ziv E, Zaida A, Salatino P, et al. Diagnostics of carbon gasification by Raman microprobe spectroscopy [J]. Proceedings of the combustion institute, 2000, 28 (2): 2369-2374.

[31] Zaida A, Bar-Ziv E, Radovic L R, et al. Further development of Raman microprobe spectroscopy for characterization of char reactivity [J]. Proceedings of the Combustion Institute, 2007, 31 (2): 1881-1887.

[32] Barbanera M, Cotana F, Di Matteo U. Co-combustion performance and kinetic study of solid digestate with gasification biochar [J]. Renewable Energy, 2018, 121: 597-605.

[33] Grigore M, Sakurovs R, French D, et al. Properties and CO_2 reactivity of the inert and reactive maceral-derived components in cokes [J]. International Journal of Coal Geology, 2012, 98: 1-9.

[34] Arrhenius. The viscosity of aqueous mixture [J]. Zeitschrift Fur Physikalische Chemie-interna-tional Journal of Research in Physical Chemistryand Chemical Physics, 1887 (1) : 285-298.

[35] Hao Jinlong. Influence of MgO/Al_2O_3 on properties of BF slag in Nanjing Iron & Steel Co., Ltd. [D]. Chongqing: Chongqing University, 2014.

[36] Zhang G H, Chou K C, Mills K. Modelling viscosities of $CaO-MgO-Al_2O_3-SiO_2$ molten slags [J]. ISIJ International, 2012, 52 (3): 355-362.

[37] Wang Xiaoliu. Iron and steel metallurgy, ironmaking part [M]. Beijing: Metallurgical Industry Press, 2000.

[38] Ram A-H. Computational analysis of modern blast furnace processes [M]. Wang Xiaoliu, translate. Beijing: Metallurgical Industry Press, 1987.

Chapter 5 Basic Research and Industrial Application of Hydrogen-Rich Fuel Injection into Blast Furnace

5.1 Brief introduction of blast furnace injection of hydrogen-rich fuel

Blast furnace ironmaking is a smelting technology that completely relies on carbon as a reducing agent, which undoubtedly requires a large amount of high-quality carbon reducing agent (coke). The shortage of coking coal resources keeps the price of coke high. Therefore, it is of great significance to inject fuel through the blast furnace tuyere as a reducing agent to replace coke. As the most active reducing agent, hydrogen has a higher reduction efficiency and reduction rate than carbon, and the reduction potential of hydrogen is 14 times that of carbon monoxide. Hydrogen energy is transportable, storable, and renewable, and its large-scale preparation technology is expected to be realized. The final product of hydrogen as a reducing agent is water, which can achieve zero emission of carbon dioxide.

At present, large-scale hydrogen production still mainly relies on fossil fuels, such as the cracking conversion, oxidation and coal gasification conversion of petroleum fuels; the other is water electrolysis hydrogen production. Both of the above two methods have the problem of CO_2 emissions, and the use of energetic gas in iron and steel enterprises to produce hydrogen can provide hydrogen sources for hydrogen metallurgy and reduce CO_2 emissions, which is conducive to energy conservation and emission reduction. Nowadays the hydrogen carried by the pulverized coal injection in the blast furnace undertakes part of the reduction task. If the blast furnace can inject substances with higher hydrogen content, the effect of reducing CO_2 emissions will be more obvious. This book mainly introduces the injection of hydrogen-rich reducing gas (such as natural gas and coke oven gas) into blast furnace based on hydrogen metallurgy theory and its influence on the performance of blast furnace raw materials. Therefore, the technology of replacing carbon with hydrogen-rich fuel as a reducing agent is expected to bring hope to the sustainable development of the steel industry. Hydrogen-rich fuel injection has attracted widespread attention from all over the world.

Natural gas is one of the safer gases. It does not contain carbon monoxide and is lighter than air. Once the gas leuks, it will spread upward immediately, and it is not easy to accumulate to form explosive gas, which is relatively safe. The use of natural gas as energy can reduce the amount of coal and oil used, thus greatly improving the environmental pollution problem: as a clean energy source, natural gas can reduce sulfur dioxide and dust emissions by nearly 100%, reduce CO_2 emissions by 60% and nitrogen oxide emissions 50%, and help to reduce the formation of acid rain, relieve the global greenhouse effect, and fundamentally improve the quality of the environment.

Coke oven gas refers to a combustible gas produced by the coke oven while producing coke and tar products when preparing coking coal. It is a by-product of the coking industry. Due to the advantages of high calorific value, high combustible components, high hydrogen content, and fast combustion, coke oven gas can be utilized in various ways. Coke oven gas is often used as a high calorific value fuel in iron and steel enterprises for maintenance and baking, heating of steel rolling furnaces, etc. However, with the improvement of energy utilization in enterprises and the use of alternative fuels, the amount of coke oven gas required for heating will keep decreasing. At the same time, coke oven gas can be used as an excellent reducing gas and chemical raw material because of its large amount of hydrogen components. Burning coke oven gas as fuel gas is actually a huge waste of energy. With the different coking coal ratio and coking process parameters, the composition of coke oven gas changes slightly. Coke oven gas generally contains 54%-59% of H_2, 5.5%-7.0% of CO and methane 24%-28%. The calorific value (standard state) of hydrogen as a fuel is only 10.8MJ/m^3, but as a reducing agent instead of CO, it can have a heat equivalent to (standard state) 12.6MJ/m^3. Therefore, blast furnace injection of coke oven gas can give full play to its role as a hydrogen-based reducing agent, making it reasonably and effectively utilized.

The current ironmaking technology tends to be stable. In order to further reduce energy consumption and realize low-carbon ironmaking, ironworkers need to actively strengthen theoretical research on hydrogen metallurgy, so as to realize the use of hydrogen-rich for blast furnace injection. Natural gas is a high-quality high-hydrogen resource. Blast furnaces in Italy, the Soviet Union, North America and other regions have all implemented natural gas injection at the blast furnace tuyere. Especially from the late 1980s to the early 1990s, 112 of the 133 blast furnaces in the Soviet Union were injected with natural gas. In the 1990s, the amount of natural gas injection into blast furnaces in North America increased significantly. The general injection volume is 40-110kg/tHM, the highest is 155kg/tHM. The No. 2 blast furnace (5,000m^3) of Japan's JFE Keihin Plant began to inject natural gas at 50kg/tHM in December 2004, and the monthly average utilization factor of the blast furnace from 2006 to 2008 reached 2.56t/($m^3 \cdot$ d), setting a new world record for a large blast furnace over 5,000m^3. Due to limited natural gas resources, high prices, and relatively concentrated distribution of production areas, only some blast furnaces in North America, Russia, and Ukraine are currently injecting natural gas, and blast furnaces are rarely injected in other regions. In recent years, hydrogen metallurgy has attracted the attention of metallurgists at home and abroad. One of the methods is to inject hydrogen-rich gas or hydrogen gas into the blast furnace. Studies have shown that the injection of natural gas into blast furnaces is beneficial to accelerate the reduction of blast furnace burden, reduce coke ratio, reduce CO_2 emissions, and is one of the means to achieve low-carbon and ultra-high efficiency of blast furnaces.

5.2 Mixed injection of natural gas and pulverized coal into blast furnace

5.2.1 Overview of natural gas injection into blast furnace

Steel is an important and irreplaceable material for the construction of the national economy. However, the steel industry is a major energy consumer, and currently, China's steel consumption accounts for about 11% of the country's total energy consumption[1]. Blast furnace ironmaking is currently and even in the coming decades the main means of obtaining pig iron from steel smelting. The energy consumption of ironmaking production, including various processes such as sintering, pelletizing, coking, and blast furnace, accounts for 73.5% of the total energy consumption of steel production processes. The energy used by steel enterprises includes coking coal, steam coal, fuel oil, and natural gas. Blast furnace injection of auxiliary fuel is a technical operation that uses coal, natural gas, or other fuels with abundant reserves to partially replace expensive coke for hot metal production. Some scholars' research shows that the energy consumption and cost of blast furnace ironmaking can simultaneously reduce the gas with the increase of coal ratio within a certain range[2]. For a long time, blast furnace ironmaking mainly relies on the injection of bituminous coal and anthracite[3]. However, some countries, such as Russia, also use natural gas as the injection material for blast furnaces because of their unique natural gas resources.

Among various fuels, gas fuel combustion is the easiest to control and has high thermal efficiency, making it a popular fuel in iron and steel plants. The main component of natural gas is CH_4 (over 90%), and hydrocarbon gas has a high calorific value. After conversion, it can obtain reducing gases mainly composed of H_2 and CO, which can be used for iron ore reduction, blast furnace injection, and direct reduction of iron ore. It is the most popular type of gas fuel[4].

The Soviet Union first injected natural gas into the blast furnace of the Petrovsky factory in 1957, achieving good results. Since then, the process of injecting natural gas into blast furnaces has been rapidly promoted worldwide. Italsider company in Italy conducted experiments on natural gas injection and the use of oxygen-enriched blast in blast furnaces. The natural gas and preheated blast were injected together from the blast furnace tuyere, causing the company's coke consumption ratio to gradually decrease. In 1967, the coke consumption ratio of Itasside company was 500kg/t, and the injection rate of natural gas was 10m^3/t. By 1971, based on the price difference between Italian coke and natural gas at that time, Itasside company found that the coke consumption ratio was 470kg/t, and the injection rate of natural gas was 35m^3/t, resulting in the best economic effect. Before 1987, the most widely used country for injecting natural gas was the Soviet Union, followed by the United States[5].

In the 1980s, due to the large-scale extraction, effective transportation, and relatively stable prices of natural gas in the world, a considerable number of countries such as the United States, the United Kingdom, France and other countries used the process of injecting natural gas for blast furnace ironmaking. At that time, the main fuel used for blast furnace ironmaking in the Iron and steel plants of Japan was high-quality heavy oil, which also used natural gas.

Direct injection of natural gas into the furnace absorbs heat and cracks into reducing gas. To prevent phenomena such as cold gas entering the furnace to cool down and incomplete natural gas reaction reducing furnace temperature, corresponding process conditions such as high oxygen enrichment rate and high blast temperature are required. In order to improve the fuel utilization and thermal efficiency of ironmaking blast furnaces, and reduce the cost of subsequent processes such as desulfurization outside the steelmaking furnace, the process of injecting high-temperature reducing gas into the furnace shaft has also been developed abroad. This process involves first decomposing hydrocarbon fuel outside the furnace, reforming it into a high-temperature (around 1,000℃) and highly reducing gas. And then injecting the gas through the indirectly reducing and intense reaction zone of the bosh or lower part of the furnace shaft into the blast furnace reduces the heat expenditure in the high-temperature zone and significantly reduces the fuel consumption of the blast furnace. The injection of reducing gas converted from natural gas ($150m^3/t$) at high temperatures into foreign ironmaking blast furnaces has reduced the coke ratio to below 300kg/t and increased the utilization coefficient of the blast furnace to over 2.4[6].

This chapter will take the natural gas and pulverized coal co-injection technology of Novolipetsk Steel in Russia as the background, and the operating technical parameters and various smelting indicators of major domestic steel plants are compared, in order to fully understand the characteristics of NLMK steel plant in Russia for mixed injection of natural gas and coal. The effective volume of the No.5 and No.7 blast furnaces of NLMK is $3,200m^3$ and $4,291m^3$, respectively. Compared to the same volume blast furnaces in China, blast furnaces of NLMK have a relatively high effective volume utilization coefficient. Because No.5 and No.7 inject $74m^3/t$ and $57m^3/t$ of natural gas while injecting coal, the coal injection rates for No.5 and No.7 blast furnaces are 112kg/t and 124kg/t, respectively, which is much lower than the average coal injection rate for blast furnaces of the same volume in China.

In order to fully compare the similarities and differences between coal injection and mixed injection of natural gas and coal, this book conducts a survey on blast furnaces of the same level in China and will compare and analyze the parameters of 11 domestic blast furnaces with an effective volume of over $3,000m^3$ and two blast furnaces of NLMK (see Fig. 5-1).

5.2.2 Impact of mixed injection of natural gas and coal on the utilization coefficient of blast furnace

The effective volume utilization coefficient of blast furnace reflects the technical and economic indicators of blast furnace production technology and management level, as well as burden and fuel conditions[7]. It is represented by the average tonnage of qualified pig iron produced per cubic meter of blast furnace effective volume per day during the specified working hours (i.e. calendar time minus major and medium repair time). The effective volume utilization coefficients of the two blast furnaces in NLMK are $2.42t/(m^3 \cdot d)$ and $2.58t/(m^3 \cdot d)$,

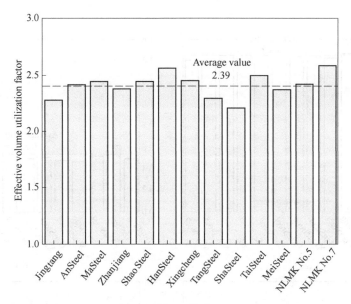

Fig. 5-1 Effective volume utilization coefficients of different blast furnaces

respectively, which are higher than the average value of $2.39t/(m^3 \cdot d)$ for blast furnaces of the same level. This indicates that mixing natural gas and coal injection into the blast furnace can effectively improve the effective volume utilization coefficient of the blast furnace. This is because natural gas is rich in hydrogen, which accelerates the reduction of iron ore in the blast furnace and increases the production of molten iron in the blast furnace.

5.2.3 Effect of mixed injection of natural gas and coal on slag performance

From Fig. 5-2, it can be seen that under high effective volume utilization coefficient, the comprehensive grades of the two blast furnaces in Russia are respectively 55.45% and 57%, lower than the average value of 58.89% for domestic blast furnaces of the same grade, while the magnesium aluminum ratio and slag content are higher than those of domestic blast furnaces of the same grade. This indicates that injecting natural gas has lower requirements for burden and fuel conditions, which is beneficial for reducing the cost of ore procurement for blast furnaces. However, a low grade of ore entering the furnace means a large amount of slag and high fuel consumption. From Fig. 5-2, it can be seen that the two blast furnaces in Russia do have a large slag volume, and the basicity of the slag is low, but the coke ratio is similar to that of the same grade of blast furnaces in China. Due to the injection of natural gas, the coal ratio injected into the blast furnace is much lower than that of the same grade of blast furnaces in China. The above phenomenon indicates that the mixed injection of natural gas and coal into the blast furnace reduces the requirement for the ore grade while improving the effective volume utilization coefficient of the blast furnace. Although the slag content of the blast furnace increases and the basicity of the slag decreases, the coke ratio of the blast furnace is basically not affected, and the coal ratio is significantly reduced.

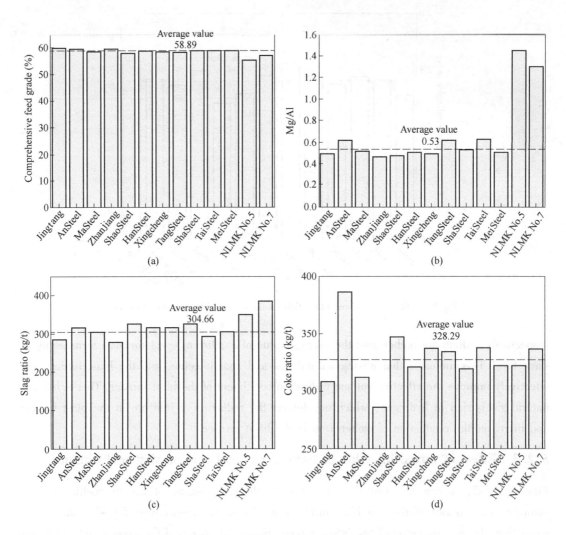

Fig. 5-2 Statistical results of different blast furnaces and fuel conditions

5.2.4 Effect of mixed injection of natural gas and coal on operating parameters of blast furnace

From Fig. 5-3, it can be seen that the blast pressure and hot blast temperature of Russian blast furnaces with the mixed injection of natural gas and coal are significantly lower than those of domestic blast furnaces of the same grade. However, the pressure difference and gas utilization rate inside the furnace are basically the same as in China. This indicates that the mixed injection of natural gas and coal into the blast furnace has lower requirements for blast operation, but it will not affect the permeability of the blast furnace and the utilization rate of gas. This result confirms the view that injecting natural gas requires lower operational requirements than injecting coal.

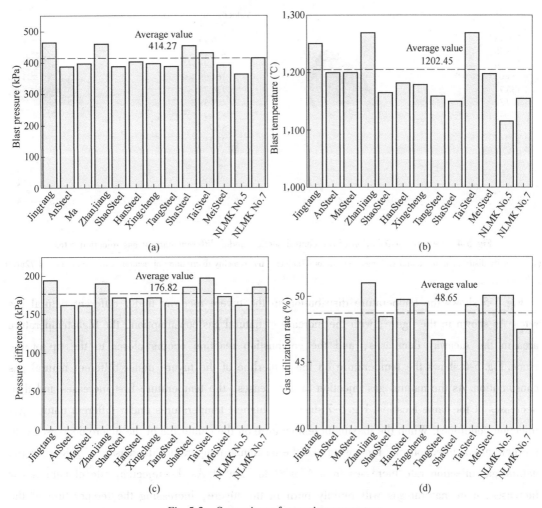

Fig. 5-3 Comparison of operating parameters

5.2.5 Impact of natural gas injection rate on the coal combustion process

Under the condition that other operating conditions remain unchanged, only the natural gas injection rate is changed to study the influence of the change of natural gas injection rate on the pulverized coal combustion process. Under the conditions of natural gas injection rates of $47m^3/t$, $52m^3/t$, $57m^3/t$, $62m^3/t$, $67m^3/t$, and $72m^3/t$, the flow and combustion characteristics of coal in the lower part of the blast furnace are simulated and studied. Fig. 5-4 shows the velocity distribution under different natural gas injection rates.

By comparing the velocity distribution in Fig. 5-4, it can be seen that the difference in velocity distribution is not significant. As the rate of natural gas injection increases, the velocity of the gas along the coal plume increases. As the rate of natural gas injection increases, more natural gas is burned in the tuyere, promoting the release of volatile, increasing the volume of gas, and promoting an increase in gas velocity.

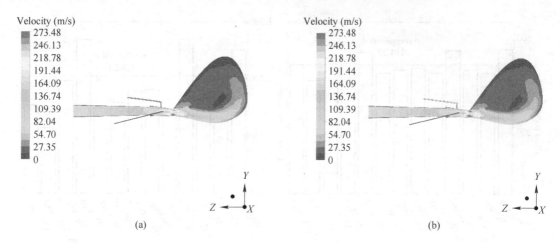

Fig. 5-4 Velocity distribution of the central section under different natural gas injection rates
(a) volocity distribution of natural gas injection rate of 47m³/t; (b) veloclity distribution of natural gas injection rate of 72m³/t

Fig. 5-5 shows the temperature distribution in the tuyere raceway under different natural gas rates. As shown in the figure, with the increase of natural gas consumption, the high-temperature area in the raceway decreases, and the combustion reaction occurs closer to the tip of the lance. Fig. 5-6 shows the temperature on the centerline of the tuyere under different natural gas consumption. As the natural gas injection rate increases, the temperature first increases and then decreases at the same location. Fig. 5-7 shows the highest temperatures under different natural gas injection conditions. When the maximum temperature on the centerline of the tuyere decreases with the natural gas injection rate increasing and the maximum temperature decreases by 32K when the natural gas injection rate increases from 47m³/t to 72m³/t. As the injection rate of natural gas increases, more natural gas will rapidly burn in the tuyere, increasing the temperature of the gas. However, the consumption of oxygen by natural gas combustion leads to a decrease in the mole fraction of oxygen, which inhibits the combustion of coal and leads to a decrease in the

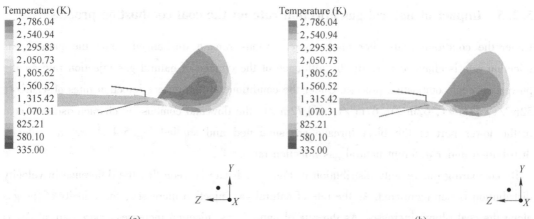

Fig. 5-5 Temperature distribution of the central section under different natural gas injection rates
(a) temperature distribution of natural gas injection rate of 47m³/t;
(b) temperature distribution at natural gas injection rate of 72m³/t

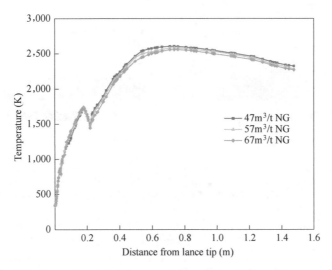

Fig. 5-6 Temperature variation curve along the centerline of tuyere under different natural gas injection rates

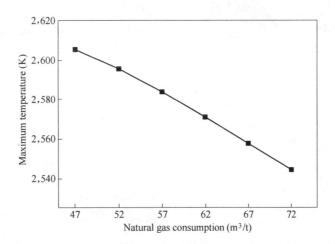

Fig. 5-7 Maximum temperature under different natural gas injection rates

temperature in the raceway. Therefore, the temperature in the raceway decreases as the natural gas injection rate increases.

Fig. 5-8 shows the gas distribution in the tuyere and raceway under different natural gas rates. Fig. 5-9 shows the mole fraction curves of O_2, CO_2, CO, H_2O and H_2 at different locations under different natural gas injection rates. From the figure, it can be seen that the distribution characteristics of gas components are similar when the injection rate of natural gas changes. As the distance from the lance outlet increases, the O_2 content gradually decreases, and the CO_2 and H_2O content first increases and then decreases. And CO and H_2 contents gradually increase with low oxygen content.

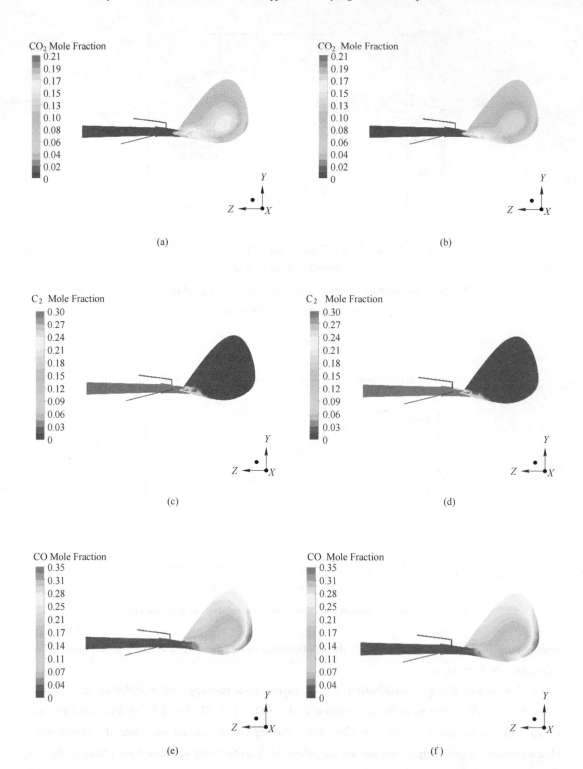

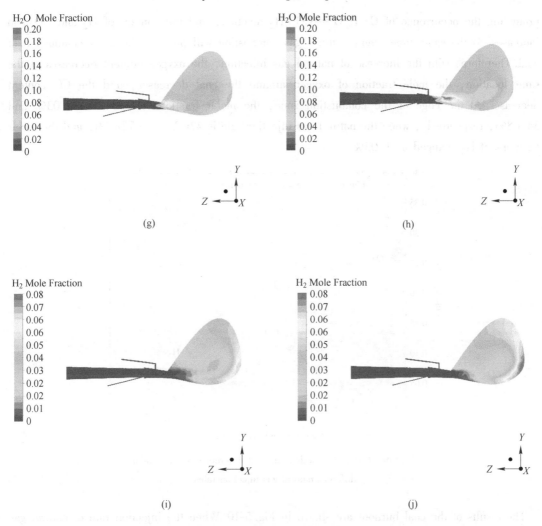

Fig. 5-8 Distribution of O_2, CO_2, CO, H_2O, H_2 at different natural gas injection rates

(a) CO_2 distribution at a natural gas injection rate of $47m^3/t$;
(b) CO_2 distribution at a natural gas injection rate of $72m^3/t$;
(c) O_2 distribution at a natural gas injection rate of $47m^3/t$;
(d) O_2 distribution at a natural gas injection rate of $72m^3/t$;
(e) CO distribution at a natural gas injection rate of $47m^3/t$;
(f) CO distribution at a natural gas injection rate of $72m^3/t$;
(g) CO distribution at a natural gas injection rate of $47m^3/t$;
(h) CO distribution at a natural gas injection rate of $72m^3/t$;
(i) CO distribution at a natural gas injection rate of $47m^3/t$;
(j) CO distribution at a natural gas injection rate of $72m^3/t$

From Fig. 5-9, it can be seen that the rate of natural gas injection has little effect on the gas composition in the furnace. When the rate of natural gas injection increases, the CH_4 content that undergoes combustion reactions in the tuyere and raceway increases, generating more H_2O, promoting the occurrence of $C+H_2O = CO+H_2$ reaction, and the content of H_2 and CO also increases. At the same time, more natural gas combustion will promote the early combustion of coal. Therefore, with the increase of natural gas injection, the oxygen content decreases at the same location, the mole fraction of oxygen around the coal decreases, and the CO content decreases. At the edge of the combustion zone, the molar fractions of CO are 35.03% and 34.08%, respectively, when the natural gas injection rate is $47m^3/t$ and $72m^3/t$, and the molar fractions of H_2 changed by 1.03%.

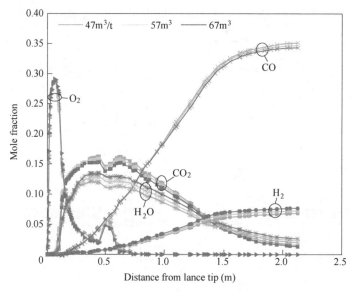

Fig. 5-9 Gas mole fraction curve at different positions under different natural gas injection rates

The results of the coal burnout are shown in Fig. 5-10. When the injection rate of natural gas increases, the burnout of coal particles at the boundary of the raceway is similar. At the boundary of the raceway, when the natural gas injection rate increases from $47m^3/t$ to $72m^3/t$, the burnouts of coal are 70.11% and 69.78%, 69.34%, 68.85%, 68.31%, 67.74%, respectively. It can be seen that only changing the injection rate of natural gas has a less significant impact on the combustion of coal.

5.2.6 Impact of natural gas injection rate and coal ratio on the coal combustion process

NLMK simultaneously uses natural gas and coal as fuel injection materials. Therefore, the effects of changes in coal and natural gas injection rates on temperature and gas composition in the raceway are studied. The changes in the lower part of the blast furnace are studied when the coal ratio is 181kg/t, the natural gas rate is $14m^3/t$, coal ratio is 167kg/t, the natural gas is $28m^3/t$, coal ratio is 153kg/t, natural gas is $42m^3/t$, coal ratio is 139kg/t, natural gas is $57m^3/t$, and coal ratio is 124kg/t.

5.2 Mixed injection of natural gas and pulverized coal into blast furnace

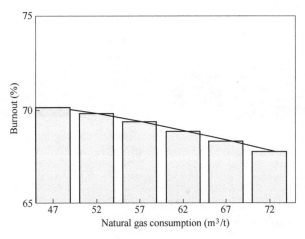

Fig. 5-10 Pulverized coal burnout at the exit of the raceway

Fig. 5-11 shows the gas distribution as fuel consumption changes and Fig. 5-12 shows the gas mole fraction curves of O_2, CO_2, CO, H_2O, and H_2 at different positions under different fuel

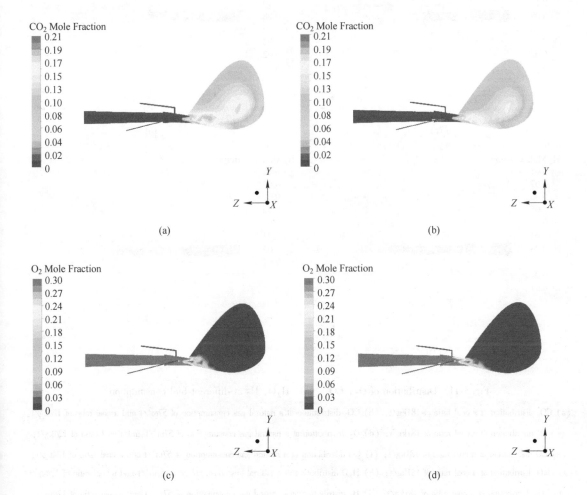

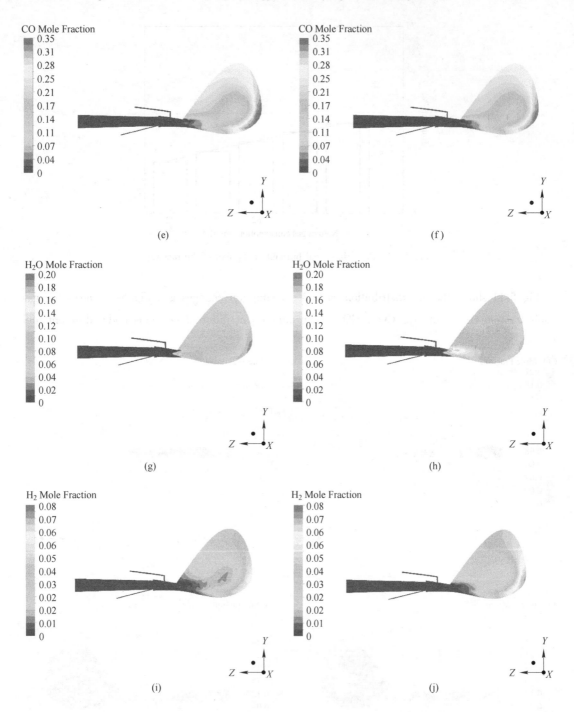

Fig. 5-11　Distribution of O_2, CO_2, CO, H_2O, H_2 at different fuel consumption

(a) CO_2 distribution at a coal ratio of 181kg/t; (b) CO_2 distribution at a natural gas consumption of $57m^3/t$ and a coal ratio of 124kg/t;
(c) O_2 distribution at a coal ratio of 181kg/t; (d) O_2 distribution at a natural gas consumption of $57m^3/t$ and a coal ratio of 124kg/t;
(e) CO distribution at a coal ratio of 181kg/t; (f) CO distribution at a natural gas consumption of $57m^3/t$ and a coal ratio of 124kg/t;
(g) H_2O distribution at a coal ratio of 181kg/t; (h) H_2O distribution at a natural gas consumption of $57m^3/t$ and a coal ratio of 124kg/t;
(i) H_2 distribution at a coal ratio of 181kg/t; (j) H_2 distribution at a natural gas consumption of $57m^3/t$ and a coal ratio of 124kg/t

consumption. From the figures, it can be seen that as the coal ratio increases and the natural gas consumption decreases, the consumption rate of O_2 slightly accelerates as the distance from the outlet of the gas outlet increases, and the consumption increases. The content of CO_2 in the upstream of the raceway increases, while the content of H_2O decreases. When the coal ratio decreases and the natural gas consumption increases, more H_2O will be generated in the tuyere and raceway, promoting the occurrence reaction of $C+H_2O = CO+H_2$, and the mole fraction of H_2 will also increase, while the mole fraction of CO will decrease.

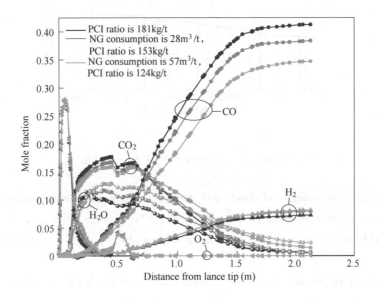

Fig. 5-12   Molar fraction of gas at different locations under different fuel consumption

The results of the pulverized coal burnout rate are shown in Fig. 5-13. When the coal ratio increases and the natural gas consumption decreases, the burnout rate at the boundary of the raceway decreases. At the boundary of the raceway, when the coal ratio is 181kg/t (No. 1), the natural gas rate is 14m³/t, the coal ratio is 167kg/t (No. 2), the natural gas is 28m³/t, coal ratio is 153kg/t (No. 3), natural gas is 42m³/t, coal ratio is 139kg/t (No. 4), natural gas is 57m³/t, and coal ratio is 124kg/t (No. 5), the pulverized coal burnout rate is 67.94%, 68.39%, 68.75%, 69.05%, 69.34%. It can be seen that changing the coal ratio and natural gas consumption simultaneously has a more significant impact on pulverized coal combustion.

5.2.7 The effect of mixed injection of natural gas and pulverized coal into blast furnace on the volume of gas in the bosh

After the mixed injection of natural gas and pulverized coal (NG/PC) into the blast furnace, it will have a series of impacts on the smelting process of the blast furnace. Therefore, it is necessary to study the impact of mixed injection of NG/PC on the smelting parameters of the blast furnace, providing a basis for the design of new blast furnace smelting processes. This

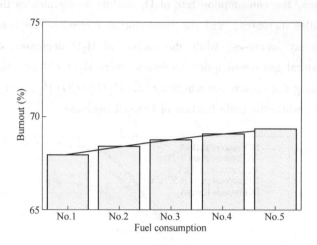

Fig. 5-13　Pulverized coal burnout rate at the exit of the raceway

section will study the effects of bosh gas volume, theoretical combustion temperature, unburned carbon residue content on slag viscosity, blast energy, tuyere and raceway, and pulverized coal burnout rate to clarify the changes in smelting parameters of mixed injection of natural gas and coal in blast furnaces.

Adopting the internationally commonly used definition of the bosh gas volume, that is, the gas generated in front of the tuyere leaves the raceway and enters the furnace bosh. Therefore, the volume of gas generated in the bosh is the volume of gas generated by combustion in the raceway. As shown in Fig. 5-14, after increasing the injection ratio of coal and decreasing the injection rate of natural gas into the blast furnace, the gas volume in the bosh decreases. The main reasons for the change in gas volume in the bosh are the following two aspects: (1) the increase in pulverized coal injection ratio /decrease in natural gas injection rate results in a decrease in the combustion heat of coal, leading to a decrease in the amount of carbon burned in front of the tuyere, resulting in a decrease in CO generated during combustion and a decrease in the volume of gas in the furnace bosh; (2) the increase in pulverized coal injection ratio/decrease in the natural gas injection rate will also lead to a decrease in the volume of gas in the furnace bosh due to a decrease in the blast volume per ton of iron. The bosh gas volume index of Chinese blast furnaces is generally controlled between 58 and 66. When the bosh gas volume index is higher than 66, it will be unfavorable for the smelting of the blast furnace. The bosh gas volume index of two NLMK blast furnaces is 65 and 58, respectively, within a reasonable range. As the injection of coal ratio increases/natural gas rate decreases, the bosh gas volume index gradually decreases, indicating that the bosh gas volume index is not a limiting indicator for NLMK blast furnace coal injection.

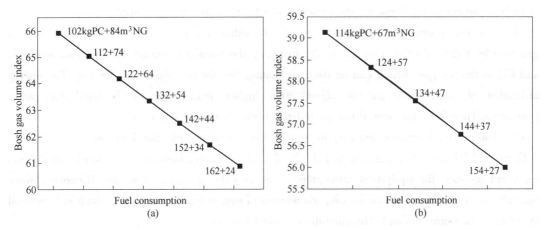

Fig. 5-14 Influence of mixed NG/PC injection on gas volume in the furnace bosh
(a) No. 5 BF; (b) No. 7 BF

5.3 Injection of coke oven gas into blast furnace

5.3.1 Significance and advantages of coke oven gas injection into blast furnace

As a by-product of coking production, coke oven gas is currently mainly used in general heating, gas-fired power generation, production of direct reduced iron and hydrogen production. Blast furnace injection of coke oven gas has been carried out a lot of fruitful research at home and abroad. In 2002, voestalpine LINZ factory injected coke oven gas at No. 5 and No. 6 blast furnaces at the same time, and achieved the achievement of injecting $125m^3$ coke oven gas per ton of iron and displacing 50kg of heavy oil. As a major producer of steel and coke in the world, China's annual coke production and consumption account for more than 50% of the world's total production. In 2007, a total of 335 million tons of coke were produced. Calculated on the basis of $420m^3$ coke oven gas per ton of coke, a total of 140.7 billion cubic meters of coke oven gas was produced. The coke oven gas was injected into the blast furnace to replace part of the expensive coke. Great powers are of great significance. Therefore, it is necessary to study the changes in smelting laws after blast furnace injection of coke oven gas, in order to enrich ironmaking theory and scientifically guide production

Blast furnace injection of coke oven gas has the following advantages:

(1) Provide good reducing agent with high hydrogen content for blast furnace. The main component of coke oven gas is hydrogen. Compared with C and CO, hydrogen is a high-quality reducing agent, which has the advantages of less heat consumption and fast reduction speed. It is injected into the blast furnace to replace the carbon in coke, which is beneficial to realize coke saving.

(2) Realize CO_2 emission reduction. The reduction product of hydrogen in the coke oven gas in the blast furnace is H_2O. Compared with the injection of pulverized coal, due to the reduction

of carbon content in the furnace, the emission of CO_2 is significantly reduced.

(3) Improve energy utilization and increase the value of coke oven gas. After the coke oven gas completes the reduction reaction in the furnace, the remaining energy is the equilibrium H_2 and CO in the top gas, which can be used as heating fuel for hot blast stove heating. The energy utilization of coke oven gas is about 80% higher than when it is used for power generation. Therefore, the overall energy utilization rate is greatly improved.

(4) The process is mature and easy to operate. Coke oven gas injection includes pressurization, delivery and injection. Compared with pulverized coal injection, because it is mainly for the gas treatment process, the equipment investment is low and the operation is simple. Moreover, since the coke oven gas does not contain ash, the amount of slag is reduced slightly, which is beneficial to reducing pressure loss and strengthening the blast furnace.

5.3.2 Development history of coke oven gas injection into blast furnaces at home and abroad

Blast furnace injection of coke oven gas abroad has a long history and has achieved good results. In the early 1980s, the Soviet Union completed the experimental research of coke oven gas injection on several blast furnaces, and mastered the smelting technology of 1.8-2.2m^3 coke oven gas instead of 1m^3 natural gas. Reached 187m^3/t and 227m^3/t.

In the mid-1980s, the No. 2 blast furnace of the Solmer Plant in France began to inject coke oven gas, and the injection volume reached 21,000m^3/h. The replacement ratio of injected coke oven gas and coke is 0.9kg/m^3. The plant firstly purifies the self-produced coke oven gas, and then uses the bed compressor to add the coke oven gas to 0.58MPa, and injects the coke oven gas into the blast furnace through the injection device. The calorific value (standard state) of coke oven gas after treatment is 19.26-20.11MJ/m^3, coke oven gas composition and impurity content are shown in Table 5-1.

Table 5-1 Composition and impurity content of No. 2 blast furnace injection coke oven gas in Solmey, France

Composition	Value (%)	Impurities	Value
H_2	60-62	Naphthalene (mg/m^3)	100-300
CH_4	24-26	Tar (mg/m^3)	<5
C_2H_4	1.8-2.2	MH_3 (mg/m^3)	<100
C_2H_6	0.8-1.2	CN (g/m^3)	<0.5
Benzene	0.9-1.1	H_2S (g/m^3)	3
CO	5.8-6.2		
CO_2	1.5		
N_2	1.5-2.5		

Since 1994, coke oven gas has been injected into the two blast furnaces of the U. S. Steel Corporation's MON VALLEY plant. The total injection volume in 2005 was 14.16 million tons, and the injection volume was about $65m^3/t$. The plant has reported that after injecting coke oven gas, the amount of natural gas injection is reduced, the air defense combustion of coke oven gas is eliminated, energy costs are reduced, and annual savings exceed $ 6.1 million.

In order to be able to inject coke oven gas, the company has carried out necessary purification treatment of coke oven gas and carried out necessary transformation on blast furnace tuyere.

China's iron and steel enterprises have also done a lot of research on blast furnace injection of coke oven gas. As early as the 1960s, Benxi Iron and Steel Co., Ltd., Xuzhou Iron and Steel Co., Ltd. and other factories conducted experimental research on coke oven gas injection in small blast furnaces, and achieved certain results. Among them, the injection coke oven gas volume of Benxi Steel is $81.6m^3/t$, the coke ratio is reduced by 60kg/t, and the output is increased by 10%-11%.

From the above, it can be concluded that:

(1) The hydrogen content in the coke oven gas is high, and after spraying into the blast furnace, the reducing atmosphere in the furnace is improved, which is beneficial to the coking and smooth flow of the blast furnace.

(2) Blast furnace injection of coke oven gas has been produced at home and abroad, technically feasible, and the process route is mature and reliable.

(3) In actual operation, coke oven gas needs to be purified to ensure that the compressor system does not affect stable operation due to the precipitation of substances such as tar and naphthalene.

(4) Use a special spray gun to inject the sales furnace gas, make necessary modifications to the direct injecting pipe of the blast furnace, and extend the coke oven gas spray gun to the front of the small tuyere to reduce the impact of the rapid combustion of coke oven gas on the small tuyere Influence.

(5) After the furnace gas is injected into the blast furnace, the combustion of the tuyere and the reducing atmosphere in the furnace will change accordingly. The blast furnace operation should be adjusted accordingly. Because the coke oven gas burns fast and the flame is short, the air flow at the edge of the blast furnace develops and the oxidation zone at the tuyere shortens. During production, the edge load must be increased to inhibit the development of the edge air flow and protect the furnace wall.

(6) After the coke oven gas is injected into the blast furnace, the moisture content of the blast furnace gas increases, and the calorific value of the gas decreases, which affects the temperature of the hot blast.

(7) Strictly control the gas temperature at the top of the furnace within 180-250℃, and carefully operate the bag dust removal system.

5.3.3 Process flow of coke oven gas injection into blast furnace

The first step is to determine the amount of coke oven gas to be injected. Take a domestic steel company as an example, according to the company's coke oven gas surplus, determine the amount of coke oven gas injecting (standard state): $55m^3/t$ in the near future, $95m^3/t$ in the long term. Compared with foreign countries have reached $100\text{-}200m^3/t$ of injecting volume, the coke oven gas injecting volume of the company's blast furnace is low both in the near and long term, and there should be no problem for the blast furnace to accept the above injecting volume.

The process of blast furnace blasting coke oven gas is simple. Take the steel company as an example, as shown in Fig. 5-15, after the gas is purified and pressurized, it is sent to the furnace coke oven gas ring pipe of blast furnace No. 1 and blast furnace No. 2 through the storage tank and pipeline respectively. Considering the limited amount of coke oven gas injecting, at the same time, in order to ensure the uniform work of the circumference of the furnace cylinder, each blast furnace is designed with 7 tuyeres still injecting pulverized coal, injecting coke oven gas tuyeres and injecting pulverized coal tuyeres using a cross-even arrangement (i.e., choose odd or even tuyeres). 7 branch pipes are led from the coke oven gas ring pipe of each of the 2 blast furnaces, which are connected to the direct injecting pipe of the odd or even tuyeres of each blast furnace through metal pipes. The coke oven gas is sprayed into the blast furnace through the metal pipe connected to the lance of the direct injecting pipe of each blast furnace's odd or even tuyeres. The material of the lance is made of high temperature and corrosion resistant material, which is specially used for coke oven gas lance. The coke oven gas ring pipe of the furnace body is provided with a number of drainage points for regularly cleaning the pipe of tar and other impurities.

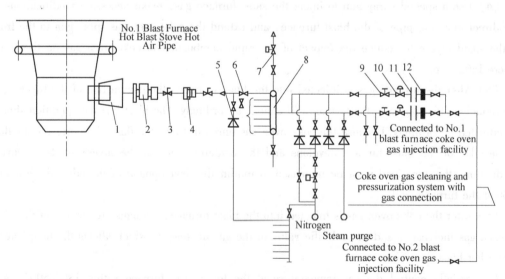

Fig. 5-15 Blast furnace coke oven gas injection process flow

1—tuyere; 2—Bullet valve; 3—ball valve; 4—quick coupling; 5—check valve; 6—shut-off valve; 7—pneumatic valve; 8—furnace ring pipe; 9—quick shut-off valve; 10—regulating valve; 11—flow orifice plate; 12—blind valve

In order to prevent the precipitation of tar and tantalum and other substances from affecting the injection, the rear pipeline of the gas storage tank is provided with 2 circuits for mutual backup, the pipeline connecting the coke oven gas ring pipe of the blast furnace body is equipped with 2 circuits for mutual backup, and there are also nitrogen and steam pipelines for security and purging. All regulating valves and quick-cut valves of the system are pneumatically controlled. The oxygen content and pressure and temperature are detected on the coke oven gas ring pipe of the blast furnace body, when the concentration is greater than 1%, the system will alarm, the quick cut valve will close and the release valve will open to release the coke oven gas in the pipe. Alarm when the pressure is lower than 0.45MPa and the temperature is lower than 60℃; when the pressure is lower than 0.35MPa, the process automatically closes the quick shut-off valve to ensure the safety of the coke oven gas injecting and opens the nitrogen purge valve for purging.

In the gas compression system, the number of compressors to be turned on is determined according to the total supply of coke oven gas. Using the control system that comes with the compressor, the outlet pressure is maintained at 0.5MPa.

Among them, the system sets complete security detection means and protection measures. It mainly includes:

(1) Pressure monitoring and safety valves for coke oven gas storage tanks and sub-gas packages.

(2) A check valve and a ball valve at the end of the lance to prevent backflow of blast furnace gas.

(3) Nitrogen gas and safety injecting steam are set for emergency injecting in the injecting main system, and nitrogen gas injecting is set in each branch pipe. Nitrogen is used for injecting of gas compression system, and steam is used for injecting of initial tank and sub-gas package.

(4) A purging and dispersion pipe is set on the gas storage tank and the gas distribution package, and the height of the dispersion pipe should be higher than the surrounding buildings. Set up drain valve and dispersion pipe in the coke oven gas ring pipe of the furnace body, and the dispersion pipe is led all the way to the dispersion valve platform on the top of the furnace.

(5) Setting up gas content monitoring and alarm system in the compressor room and coke oven gas ring pipe of the furnace body, and setting up necessary fire-fighting measures.

(6) The control system sets up the necessary safety interlock, and automatically stops or automatically closes the quick shut-off valve and purges in case of abnormal conditions.

5.3.4 Effect of coke oven gas injection on the smelting law of blast furnace

The material balance and heat balance models of blast furnace smelting are constructed to calculate the influence of coke oven gas injecting on the smelting laws and economic and technical indexes of blast furnace. Based on the production conditions of $2,000m^3$ blast furnace in a plant, where the blast temperature is 1,100℃, coal ratio is 140kg/t, blast humidity is 1%, ore into furnace grade is 56.3%, slag alkalinity is 1.15, coke oven gas injecting temperature is 40℃, where the hydrogen utilization ratio remains unchanged at 0.45, coke ratio, oxygen enrichment rate and coke oven gas injecting volume are adjustable. The specific raw fuel conditions are shown in Table 5-2 and Table 5-3.

Table 5-2 Main chemical composition of ore into the furnace   (%)

Composition	TFe	FeO	CaO	SiO$_2$	MgO	Al$_2$O$_3$	S
Sinter	54.73	8.45	11.42	6.15	2.25	2.25	0.033
Pellet	59.7	1.82	0.52	12.04	0.70765	1.0663	0.03
Lump ore	63.44	0.5	0.49	2.31	0.5	2.6	0.059

Table 5-3 Composition of fuel into furnace   (%)

Composition	FC	S	Ash content						Volatile matter	Moisture
			CaO	SiO$_2$	MgO	Al$_2$O$_3$	Others	Total		
Coke	84.95	0.65	1.21	5.21	0.34	4.71	0.16	13.23	1.17	0.00
Pulverized coal	78.38	0.65	0.93	5.39	0.36	3.53	0	11.18	8.29	1.50
Coke oven gas	H$_2$=59.8%, CH$_4$=20.5%, CO=6.5%, C$_n$H$_m$=4.0%, N$_2$=6.2%, CO$_2$=2.5%, O$_2$=0.5%									

The simulation is divided into 5 stages, where the blast temperature, coal ratio, and blast humidity are constant.

Base period: not oxygen-rich, coke ratio 330kg/t, coke oven gas injecting volume 0m^3/t, theoretical combustion temperature between 2,000 and 2,300℃, heat loss not less than 3%;

The first stage: no oxygen enrichment, coke oven gas injecting volume is 0, 40-90m^3/t, respectively, the increase is 10m^3/t, coke ratio is variable, theoretical combustion temperature is between 2,000-2,300℃, heat loss is not less than 3%;

The second stage: oxygen enrichment rate of 2%, coke oven gas injecting volume of 0, 40-120m^3/t, respectively, the increase is 10m^3/t, coke ratio is variable, theoretical combustion temperature is between 2,000-2,300℃, heat loss of not less than 3%;

The third stage: oxygen enrichment rate of 4%, coke oven gas injecting volume of 0, 40-140m^3/t, respectively, the increase is 10m^3/t, coke ratio is variable, theoretical combustion temperature is between 2,000-2,300℃, heat loss of not less than 3%;

The fourth stage: oxygen enrichment rate of 6%, coke oven gas injecting volume of 0, 40-160m^3/t, respectively, the increase is 10m^3/t, coke ratio is variable, theoretical combustion temperature is between 2,000-2,300℃, heat loss of not less than 3%.

The simulation results show that after blast furnace injecting coke oven gas has an effect on direct reduction degree, coke ratio, belly gas volume, theoretical combustion temperature and top gas volume and top gas composition, the effect results are as follows.

5.3.4.1 Changes in the direct reduction degree after oxygen-enriched coke oven gas injection

After coke oven gas injecting, as coke oven gas is a hydrogen-rich fuel, it makes the internal H_2 content of blast furnace increase, which changes the conditions of iron oxide and C gasification and promotes the indirect reduction of iron oxide inside the blast furnace and reduces the direct reduction degree.

The reasons for blast furnace injecting coke oven gas to promote indirect reduction are:

(1) The content of reducing gas in the gas composition increases, and the reduction of the blast volume after oxygen enrichment makes the nitrogen brought in by the blast decrease, and the concentration of reducing gas increases.

(2) The amount of reducing gas produced after coke oven gas injecting is greater than that produced by coke, and the absolute amount of reducing gas increases despite the decrease in coke ratio, and the amount of reducing gas per unit of pig iron increases, promoting the development of indirect reduction.

(3) The conventional blast furnace ironmaking method uses CO gas as a reducing agent to reduce oxygen in iron ore, because the molecules of CO gas are large and difficult to penetrate into the iron ore, while the molecules of H_2 gas are extremely small and can easily penetrate into the iron ore, and its penetration rate is about five times that of CO gas, and the excellent kinetic conditions also promote the indirect reduction. There is no good way to determine the degree of direct reduction inside the blast furnace, and the empirical formulae summarized by previous authors on the basis of a series of experiments are important guidance for us to conduct research.

Professor A. H. Ram of the USSR summarized the empirical formula for calculating the degree of direct reduction under different injecting conditions:

$$r_d = r_d^0 \times 10^{-s\lambda}(0.684 + 0.01t_B^{0.5})/(0.96 + 4\varphi) \tag{5-1}$$

where, r_d^0 is r_d values for the base period; t_B is blast temperature; s is the injection volume of reducing material, which unit is $m^3(kg)/kg$; φ is blast humidity, which unit is mL/m^3; λ is the coefficient of chemical composition of injecting material:

$$\lambda = 0.2(\overline{C}) + 0.9(\overline{H}) \tag{5-2}$$

where, $\overline{C}$, $\overline{H}$ are carbon and hydrogen content per unit of injecting material, respectirely, which unit is m^3/m^3 (m^3/kg).

According to the calculation formula of A. H. Ram, combined with the influence of coke oven gas and pulverized coal injecting amount, the results of the influence of coke oven gas injecting amount on the direct reduction degree are obtained, see Table 5-4.

Table 5-4 Direct reduction degree under different coke oven gas injecting volume (%)

Coke oven gas injecting volume	0	40	50	60	70	80	90
r_d	0.421	0.381	0.371	0.362	0.353	0.344	0.335

Continued Table 5-4

Coke oven gas injecting volume	100	110	120	130	140	150	160
r_d	0.327	0.319	0.311	0.303	0.295	0.288	0.281

5.3.4.2 Effect of oxygen-enriched coke oven gas injection on coke ratio

From Fig. 5-16, we can see that the coke ratio decreases with the increase of coke oven gas injecting under the same oxygen-rich condition, and the coke ratio decreases by 4-5kg/t for every 10m³/t of coke oven gas increase when the coke oven gas injecting volume is less. The four points A, B, C and D are the lowest coke ratios corresponding to different oxygen-rich conditions, and the coke ratios and coke oven gas injecting volumes are 305kg/t, 50m³/t, 299kg/t, 90m³/t, 295kg/t, 110m³/t, 291kg/t, 140m³/t respectively; the coke ratio increases rapidly when the coke oven gas injecting volume is increased again, mainly because the coke oven gas injecting will reduce the theoretical combustion temperature in the raceway, as shown in Fig. 5-16, in order to ensure the thermal state of the hearth, more coke needs to be burned, which in turn increases the coke ratio.

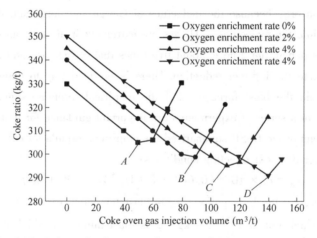

Fig. 5-16 Relationship between coke ratio and coke oven gas injection volume under different oxygen enrichment conditions

It can also be derived from Fig. 5-17 that the coke ratio increases with increasing oxygen enrichment rate for the same coke oven gas injecting volume. The increase of oxygen enrichment rate reduces the blast volume, the physical heat brought in by the blast is reduced, and more coke needs to be burned to meet the balance of heat income and expenditure of the blast furnace. Therefore, the coke ratio cannot be reduced by simply injecting coke oven gas and oxygen enrichment, and the combination of oxygen enrichment and coke oven gas injecting is needed to achieve the effect of coke ratio reduction.

Fig. 5-18 shows the relationship between the minimum coke ratio and coke oven gas injecting volume under different oxygen enrichment conditions, it can be seen that the coke ratio reaches

5.3 Injection of coke oven gas into blast furnace

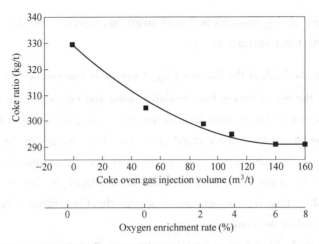

Fig. 5-17 Relationship between the lowest coke ratio and coke oven gas injection volume at different oxygen enrichment rates

the minimum value of 291kg/t when the oxygen enrichment rate is 6% and the coke oven gas injecting volume is 140m³/t, further increasing the oxygen enrichment rate and coke oven gas injecting volume not only cannot play a role in reducing the coke ratio, but also increase the oxygen production cost and fuel ratio. Fig. 5-18 shows the relationship between coke oven gas injecting and replacement ratio under different oxygen enrichment conditions, with the increase of coke oven gas injecting, the replacement ratio is decreasing, and the replacement ratio is only 0.378kg/m³ when the coke oven gas injecting is 160m³/t. Therefore, the oxygen enrichment rate of 6% and the coke oven gas injecting is 140m³/t is the best oxygen enrichment injecting, which not only reduces the coke ratio significantly, but also can make the replacement ratio and the fuel ratio of the coke oven gas injecting. Therefore, the oxygen-enriched rate of 6% and the oxygen-enriched gas injecting volume of 140m³/t is the best combination, which can not only greatly reduce the coke ratio but also keep the replacement ratio and oxygen-enriched rate in the suitable range.

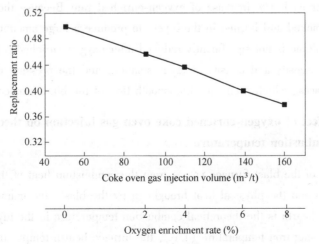

Fig. 5-18 Relationship between coke oven gas injection volume and replacement ratio

5.3.4.3 Effect of oxygen-enriched coke oven gas injection on the amount of gas in the blast furnace bosh

The amount of gas in the bosh of the furnace (V_{BG}) currently has two definitions, one is that the gas generated before the tuyere leaves the circulation zone and enters the the blast furnace bosh, so the amount of gas in the blast furnace bosh is the amount of gas formed in the tuyere raceway; the second is that the gas in the blast furnace bosh should be the gas in the drop zone, and in addition to the gas formed in the tuyere raceway, it should also take into account the amount of CO generated by partial direct reduction, etc. This book adopts the former viewpoint which is common internationally, that is, the amount of gas in the blast furnace bosh is equal to the amount of gas generated in the combustion zone.

The amount of gas in the blast furnace bosh increases and the combustion zone expands after coke oven gas blast injecting. After the coke oven gas blast furnace injecting, the hydrocarbons such as CH_4 in the gas are burned in front of the air outlet to produce a large amount of H_2, so that the amount of gas in the blast furnace bosh increases under the same oxygen-enriched rate condition, as shown in Fig. 5-19. After oxygen enrichment, the concentration of oxygen in the blast air increases, the concentration of nitrogen decreases, the blast volume required to burn 1kg of carbon decreases, and accordingly the amount of gas produced by combustion in front of the tuyere decreases, and the combustion zone shrinks. Comparing the effects on the amount of gas in the blast furnace bosh for every $10m^3/t$ of coke oven gas injetion and 1% increase of oxygen enrichment rate, it is found that the effect of increasing coke oven gas injetion is less than the effect of decreasing oxygen enrichment, so the amount of gas in the blast furnace bosh is $1,365m^3/t$ at $140m^3/t$ of coke oven gas injection and 6% of oxygen enrichment, which is basically consistent with the amount of gas in the blast furnace bosh at $1,368m^3/t$ in the base period stage.

Fig. 5-20 indicates the blast volume and furnace bosh gas volume corresponding to the lowest theoretical coke ratio under different oxygen-enriched rate conditions, and it can be seen that the blast volume decreases with the increase of oxygen-enriched rate, and the blast volume decreases by about $30m^3/t$ for each 1% increase of oxygen-enriched rate. Because the organic gas in the coke oven gas is cracked and burned in the tuyere to produce a large amount of CO and H_2, the furnace bosh gas volume is not significantly reduced after oxygen-enriched injecting of coke oven gas. However, the viscosity and density of H_2 is smaller, and the pressure difference inside the blast furnace decreases, which promoting the smooth flow of the blast furnace charge.

5.3.4.4 Effect of oxygen-enriched coke oven gas injection on theoretical combustion temperature

Almost all the heat of the blast furnace comes from the combustion heat of the fuel in the tuyere raceway of the wind and the physical heat brought in by the blast, the main sign of the thermal state of the furnace hearth is the theoretical combustion temperature in the tuyere raceway, which not only affects the slag iron temperature (i.e., the furnace hearth temperature) but also affects

5.3 Injection of coke oven gas into blast furnace

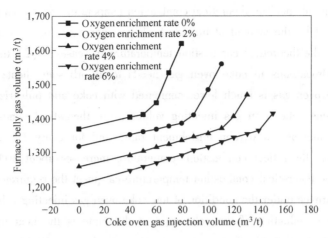

Fig. 5-19 Relationship between coke oven gas injection volume and furnace bosh gas volume

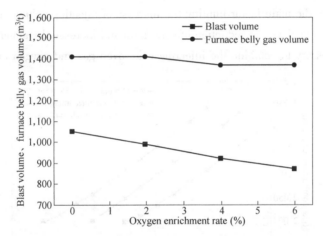

Fig. 5-20 Relationship between blast volume, furnace bosh gas volume and oxygen enrichment rate

the shape of the fusion zone, the distribution of gas flow and the reduction reaction of iron oxides, etc. The theoretical combustion temperature is too high or too low will lead to blast furnace disorder, so that blast furnace production changes, the appropriate theoretical combustion temperature should be able to meet the normal smelting of the blast furnace required heat and hearth temperature, both to ensure that the liquid slag iron fully heated, hearth heat exchange and reduction reaction is carried out normally, but also to enable the rapid combustion of injecting fuel in the tuyere raceway. Domestic blast furnace after injecting fuel, the theoretical combustion temperature is maintained between 2,000-2,300℃, in the simulation calculations, the lower limit of the theoretical combustion temperature taken 2,000℃, the upper limit of 2,300℃.

Fig. 5-21 shows the variation curve of theoretical combustion temperature with the amount of coke oven gas injecting under different oxygen enrichment rates. At the same oxygen enrichment rate, the theoretical combustion temperature decreases with the increase of coke oven gas injecting volume, and the theoretical combustion temperature decreases about 25℃ for every 10m^3/tHM increase of injecting volume. The theoretical combustion temperature decreases after coke oven gas injecting because: (1) the amount of furnace bosh gas increases, the heat needed to heat the furnace bosh gas to the theoretical combustion temperature increases; (2) the decomposition of CH_4 and other hydrocarbons in coke oven gas needs to absorb some heat; (3) the heat of combustion of coke oven gas is much lower compared with coke and pulverized coal. Under the condition of the same coke oven gas injecting volume, the theoretical combustion temperature increases with the increase of oxygen enrichment rate, and for every 1% increase of oxygen enrichment rate, the theoretical combustion temperature increases about 30℃. The calculation results show that the theoretical combustion temperature can meet the requirements of the thermal state of the blast furnace under the condition of low coke oven gas injecting volume, and the main factor limiting further reduction of the coke ratio at this time is the heat income of the blast furnace. With the increase of injecting volume, the theoretical combustion temperature decreases, and when the injecting volume increases to a certain degree, the theoretical combustion temperature becomes the main factor limiting the increase of injecting volume and the reduction of coke ratio. In order to achieve the reduction of coke ratio and increase the amount of coke oven gas injecting, it is necessary to combine the injecting coke oven gas and oxygen enrichment.

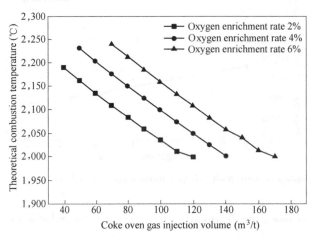

Fig. 5-21 Trend of theoretical combustion temperature with the amount of injection

5.3.4.5 Influence of oxygen-enriched coke oven gas injection on the change of gas volume and composition at the top of the furnace

After oxygen-enriched coke oven gas injection of blast furnace, with the increase of coke oven gas injection, the amount of blast furnace top gas decreases and the content of reducing gas increases. The reducing gas content in the base period gas is 21.14%, and when the coke oven gas is injected at 140m³/t and the oxygen enrichment rate is 6%, the reducing gas content increases to 29.3%, and the gas calorific value also increases from 2,600kJ/m³ to 3,600kJ/m³, as shown in Fig. 5-22. The significant increase of the calorific value of blast furnace gas broadens the utilization range of blast furnace gas, improves the utilization value of blast furnace gas and increases the enterprise benefits.

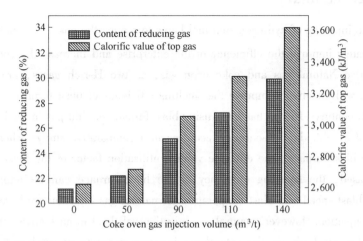

Fig. 5-22 Content of reducing gas and calorific value of top gas under different conditions of coke oven gas injection volume

After the blast furnace oxygen-enriched injection of coke oven gas, the emission of CO_2 is reduced. Because the main components of coke oven gas are H_2 and CH_4, H_2 participates in the indirect reduction reaction product of water, which replaces part of C in the blast furnace, resulting in the reduction of blast furnace CO_2 production and emissions. Compared with the CO_2 emissions in the base period, the net CO_2 emissions are reduced by 6.1% when the coke oven gas injecting volume is 140m³/t, as shown in Fig. 5-23. For an iron and steel enterprise with an annual production of 10 million tons of iron water, it can reduce CO_2 emission by 220 million m³ per year, which shows that oxygen-enriched blast furnace injecting of coke oven gas is of great significance to reduce the CO_2 emission of the enterprise, alleviate the pressure of environmental protection and improve the surrounding environment.

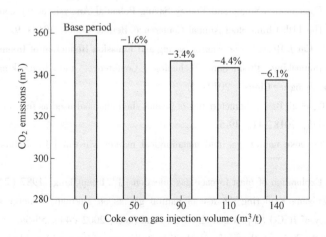

Fig. 5-23 CO_2 emissions at different coke oven gas injection volumes

5.4 Chapter summary

The blast furnace injecting of hydrogen-rich fuel helps to reduce the coke ratio and CO_2 emission, which will ultimately improve the efficiency of the enterprise and increase the competitiveness of the steel enterprise. Natural gas and coke oven gas, as two H-rich gases, are used for blast furnace injecting will effectively improve the smelting efficiency of blast furnace.

Since the 20th century, Russia has been using blast furnace natural gas in jection technology to take advantage of its abundant resources. According to a comparative study, when natural gas is injected into the blast furnace, the effective volume utilization factor of the blast furnace can be effectively increased, the smelting efficiency of the blast furnace can be improved, and the requirements of blast speed and blast temperature for natural gas injection are lower than those for pulverized coal injection. However, natural gas injection will lead to an increase in the amount of gas in the blast furnace bosh, so the natural gas injection needs to be combined with solid fuels such as pulverized coal, and the upper limit of natural gas injection needs to be clarified.

As a by-product of coke production, the use of coke oven gas for blast furnace injecting has been fruitfully studied in China and abroad. Coke oven gas, also as a hydrogen-rich fuel, can be used as a good reducing agent for blast furnace injecting to improve energy utilization and achieve CO_2 emission reduction, and its injecting process is simple. Coke oven gas injecting can significantly reduce direct reduction, develop indirect reduction, and reduce blast furnace coke ratio. At the same time, like natural gas, coke oven gas injecting also affects the amount of bosh gas and top gas composition of the blast furnace. The results of both basic research and industrial applications show that both natural gas and coke oven gas injecting can achieve energy saving and consumption reduction, but the use for different blast furnace should be determined by the condition of the blast furnace.

References

[1] Wang W. Energy Consumption Status and Energy Saving Potential Analysis of my country's Iron and Steel Industry [C] //The 11th China Steel Annual Conference. Beijing China, 2017: 9.
[2] Shen F, Zhang Q, Cai J. Research on Energy Saving and Emission Reduction of Ironmaking System [C] // 2012 National Ironmaking Production Technology Conference and Ironmaking Academic Annual Conference. Wuxi, Jiangsu, China , 2012: 5.
[3] Li C, He X, Li C, et al. Basic characteristics of peanut shell charcoal used as fuel for blast furnace injection [J]. Coal conversion, 2018, 41: 49-53.
[4] Xu Zhigang. Gas for coke natural gas and metallurgical market win-win [J]. Natural Gas Industry, 2000 (5): 86-90.
[5] Wen Guangyuan. Exploration of blast furnace gas injection [J]. Ironmaking, 1987 (2): 12-16.
[6] Zou Ming, Xu Zhigang. Exploring the use of natural gas as an alternative energy source to improve the economic efficiency of JISCO [J]. Research on Iron and Steel, 2002 (4): 55-59.
[7] Zhang Zhanpeng. Analysis of the main production indicators of ironmaking blast furnaces of key steel enterprises [J]. China Steel, 2014 (5): 25-26.

Chapter 6　Prospects for the Development of Blast Furnace Injection Fuel Resources in China

The steel industry is an important basic industry of China's national economy, which consumes a large amount of energy and mainly relies on coal resources. The distribution of China's coal resources is extremely uneven, among the proven reserves, bituminous coal accounts for 73.7%, anthracite accounts for 7.9%, lignite coal accounts for 6.8%, and other coal types account for 11.6%, and the reserves of high quality coking coal and fat coal in bituminous coal only account for 7.9%. The coking coal and fertilizer coal resources and anthracite resources reserves for blast furnace injection, on which blast furnace coke production depends, are facing great challenges. Blast furnace pulverized coal injection is an important measure for steel enterprises to alleviate the shortage of coking coal resources and reduce the coke ratio and pig iron production cost, and is the mainstream trend of the development of iron making technology in the world. However, the 100% high quality of the coal blasting can no longer meet the requirements of China's current blast furnace ironmaking on environmental protection and energy saving, sustainable development of resources and cost reduction and efficiency.

More and more iron and steel enterprises continue to expand the range of coal resources for ironmaking, the extensive use of poor lean coal, bituminous coal, lignite for blast furnace injection, try to use semi-coke for blast furnace injection, although some breakthroughs have been made, but also facing a lot of technical problems. Among them, the flammable and explosive of high volatile bituminous coal poses a serious safety problem for blast furnace injection; frequent changes in multiple coal resources lead to fluctuations in the thermal state of the blast furnace cylinder and gas flow, which seriously restrict the stability of blast furnace smelting. The main reason is the lack of fine management and scientific selection of the introduction of new coal resources in steel enterprises, the failure to establish a systematic evaluation system and process standards based on the basic theory of mass-energy transformation of coal in the ironmaking process, and the use of new fuels has not yet been matched with the development of equipment. The development of blast furnace fuel resource expansion technology in China has the following directions:

(1) Create a scientific evaluation system for coal used in blast furnace injection. At present, the significant characteristics of blast furnace injection coal is a variety of chaotic, unstable production areas, frequent changes in the material, how to scientifically select the appropriate economic coal for blast furnace injection is a prominent problem facing steel enterprises to reduce costs and increase efficiency. Therefore, it is necessary to systematically study the combustion mechanism of pulverized coal in the blast furnace windrowing area, develop a cost-effective

evaluation model of blast furnace integrating economic and technical indexes, construct a scientific evaluation system of coal resources for ironmaking, and provide theoretical guidance for the selection of economic coal types and coal injection volume for blast furnace coal injection.

(2) Accelerate the use of blast furnace injection low-grade coal technology. High volatile low rank coal blast furnace injection can not only relieve the dependence of iron making on high quality anthracite resources, but also help to improve the burning rate of mixed coal. However, the flammable and explosive properties of high volatile bituminous coal bring serious safety hazards to the blast furnace injection system; at the same time, some steel enterprises blast furnace injection system equipment is dilapidated, which restricts the application of low-rank coal in the blast furnace ironmaking. Therefore, China needs to increase the technology promotion of blast furnace injection low rank coal, formulate relevant national standards, strengthen the training of enterprise staff, improve the equipment level of iron and steel enterprises, and realize the large-scale application of high proportion low rank coal in the field of blast furnace injection as soon as possible.

(3) Enhance the targeted preparation of blast furnace injection fuel from low-rank coal. Lang char is a semi-coke of low quality coal dry-fried at medium and low temperature, which has the characteristics of high fixed carbon, high specific resistance, high chemical activity, low ash, low aluminum, low sulfur and low phosphorus. However, the poor grindability of lang char leads to serious wear of pulverizing system and blowing system; high moisture content leads to lower effective heating value of lang char. Therefore, it is necessary to develop customized production technology for iron making industry to solve the problems of high fluctuation of carbon composition and poor grindability.

(4) Accelerate the research and development and industrial application of blast furnace injection biomass technology. Biomass coke has the advantages of low ash content, good combustibility and reactivity, suitable for use as a heating agent and reducing agent in the ironmaking process, and has good economic and environmental benefits. However, only foreign countries have done small blast furnace injection experiments, the domestic research is relatively small, so the development of biomass blowing technology in China has a long way to go. In the pretreatment of biomass, blowing process route development and equipment development still need in-depth research.

(5) Research and development of blast furnace hydrogen injection and hydrogen-rich reduction gas technology. Hydrogen is the most clean and efficient energy source for iron making, and the end product of hydrogen as reductant is water, which can reach zero CO_2 emission, so hydrogen metallurgy has gained widespread attention from all over the world. Blast furnace ironmaking is the main ironmaking process in China, and the research of blast furnace injection hydrogen and hydrogen-rich reductant (coke oven gas, natural gas, etc.) technology is of great significance to China's ironmaking energy saving and emission reduction, and is the focus of development of blast furnace injection technology in China.

(6) Develop blast furnace injection solid waste technology. At present, China's steel

industry, while completing the supply of basic raw materials, also needs to shoulder the function of social solid waste elimination. The tuyere area of blast furnace has the characteristics of ultra-high temperature and high pressure, which is good for processing carbon containing waste such as waste plastics, waste tires and waste wood. Carbon containing solid waste used in blast furnace injection not only helps to reduce the consumption of fossil fuels in blast furnace iron making, but also helps to recycle solid waste and solve the secondary pollution problem brought by traditional solid waste treatment, truly realizing the transformation of waste into treasure.